79 Advances in Polymer Science

Biopolymers/ Non-Exclusion HPLC

With Contributions by
J. D. Andrade, G. Glöckner, V. Hlady,
P. Klosiński, T. E. Lipatova, S. Penczek,
N. A. Platé, L. I. Valuev

With 81 Figures and 29 Tables

Springer-Verlag Berlin Heidelberg GmbH

ISBN 978-3-662-15191-4 ISBN 978-3-540-39774-8 (eBook)
DOI 10.1007/978-3-540-39774-8

Library of Congress Catalog Card Number 61-642

Originally published by Springer-Verlag Berlin Heidelberg New York in 1986
Softcover reprint of the hardcover 1st edition 1986

2154/3020-543210

Editors

Editorial

With the publication of Vol. 51 the editors and the publisher would like to take this opportunity to thank authors and readers for their collaboration and their efforts to meet the scientific requirements of this series. We appreciate the concern of our authors for the progress of "Advances in Polymer Science" and we also welcome the advice and critical comments of our readers.

With the publication of Vol. 51 we would also like to refer to a editorial policy: *this series publishes inited, critical review articles of new developments in all areas of polymer science in English (authors may naturally also include workes of their own).* The responsible editor, that means the editor who has invited the author, discusses the scope of the review with the author on the basis of a tentative outline which the author is asked to provide. The author and editor are responsible for the scientific quality of the contribution.

Manuscripts must be submitted in content language and form satisfactory to Springer-Verlag. Figures and formulas should be reproducible. To meet the convenience of our readers, the publisher will include "volume index" which characterizes the content of the volume.

The editors and the publisher will make all efforts to publish the manuscripts as rapidly as possible, i.e., at the maximum six months after the submission of an accepted paper. Contributions from diverse areas of polymer science must occasionally be united in one volume. In such cases a "volume index" cannot meet all expectations, but will nevertheless provide more information than a mere volume number.

Starting with Vol. 51, each volume will contain a subject index.

Editors Publisher

Table of Contents

Protein Adsorption and Materials Biocompatibility: A Tutorial Review and Suggested Hypotheses

J. D. Andrade and V. Hlady
College of Engineering, University of Utah, Salt Lake City, Utah 84112/U.S.A.

A comprehensive review of protein adsorption at solid-liquid interfaces is presented, including a brief review of protein structure and the principles of protein adsorption. Adsorption-based biocompatibility hypotheses and correlations are discussed, including surface charge, interface energetics, passivation, protein-resistant surfaces, and the role of adsorbed immunoglobulins and complement. New methods for the study of protein adsorption are discussed, including total internal reflection techniques (absorbance, fluorescence, and Raman) and ellipsometry. Qualitative "rules of thumb" of protein adsorption are also presented.

Advances in Polymer Science 79

1 First Observable Event?

It is commonly stated that the first readily observable event at the interface between a material and a biological fluid is protein or macromolecule adsorption. Clearly other interactions precede protein adsorption: water adsorption and possibly absorption (hydration effects), ion bonding and electrical double layer formation, and the adsorption and absorption of low molecular weight solutes — such as amino acids. The protein adsorption event must result in major perturbation of the interfacial boundary layer which initially consists of water, ions, and other solutes.

Body fluids have a rich and complex composition, permitting a wide range of interactions and competitive processes.

The great majority of the available work has focused on blood/plasma/serum applications. In recent years the interaction of tear components with contact lenses have become very important.

Much of the early work is reviewed by Brash and Lyman [1] and by Vroman [2]. Protein adsorption is a very old field — the earliest studies probably being monomolecular film observations using Langmuir-Blodgett troughs [3]. A comprehensive monograph which reviews protein adsorption is now available [4].

Early observations showed that blood clotted faster in clean glass tubes than in siliconized glass. When Mr. Hageman's blood failed to clot *in vitro*, it was speculated that a "Hageman Factor" was responsible for the *in vitro* "activation" of blood coagulation (see Ref. [2] for a delightful account).

Early observations, particularly the work of Vroman [2,5], showed that proteins adsorb on practically any surface. Vroman first showed [5a] that adsorption of protein on a hydrophilic surface rendered it less hydrophilic and also that protein adsorption on a hydrophobic surface rendered it hydrophilic.

Knowledge and understanding of protein structure and properties in the 1950's was rapidly evolving. The unique secondary, tertiary, and even quaternary structures of proteins were becoming understood [6–8] and the delicateness of protein three-dimensional conformation was recognized, including the possibility for "denaturation" at liquid/air and solid/liquid interfaces [2,3].

The fact that Hageman Factor, a plasma protein, circulated free in blood without inducing coagulation, but when exposed to glass or silicate clay surfaces did induce coagulation, led to a unique and novel hypothesis: changes in conformation induced by the adsorption of Hageman Factor result in enzymatic activity which initiates a sequence of reactions leading to fibrin formation. Although interface-induced denaturation of proteins at liquid/air interfaces was well-known, the interest in *in vitro* blood handling and the early development of medical devices provided an impetus for understanding and controlling surface-induced coagulation.

At about the same time, the field of protein separation and purification was undergoing rapid development. The introduction of materials for protein chromatography, such as cross-linked dextran, agarose, and polyacrylamide, provided a means to study protein-surface interactions, as well as to dramatically advance knowledge in protein biochemistry.

2 Protein Structure

The treatment and understanding of protein adsorption requires familiarity with modern concepts of protein structure and function, such as provided in Ref. [6-8]. A concise review is available in Chapter 1 of Ref. [4].

Proteins are biological macromolecules constructed for specific and unique functions. They are high molecular weight polyamides produced by the specific co-polymerization of up to 20 different amino acids. The amino acid composition, called the primary structure, is generally unique and specific to each particular protein. The hydrogen bonding characteristics of the polyamide bond in the backbone of proteins result in various secondary structures, such as the well-known α-helix and β-sheet. Intramolecular associations, including ionic interactions, salt bridges, hydrophobic interactions, hydrogen bonding, and covalent disulfide bonds, result in a unique tertiary structure for each polypeptide chain. Finally, two or more polypeptide chains, each with its own primary, secondary, and tertiary structure, can associate to form a multi-chain quaternary structure. Most proteins contain short, carbohydrate sequences off the main polypeptide chain and are therefore called glycoproteins.

The fundamental principles of protein structure and function are available in modern biochemistry textbooks [6-7]. The current edition of Stryer is particularly recommended [7]. The books by Dickerson and Geis [6, 9] are excellent tutorials on the fundamental principles of protein structure. The modern text by Schulz and Schirmer [8], *Principles of Protein Structure*, is outstanding. In addition, modern physical

Table 1. Minimal protein information required for adsorption studies

Quantity	Method	Significance/Information
Molecular Weight	Light scattering, Osmometry, Gel permeation chromatography	Size, shape, mass of molecule.
Size & Shape	Light scattering sedimentation	Size, shape, mass of molecule.
Amino Acid Composition	Amino acid analysis	Basic.
Electrophoretic Mobility	Gel electrophoresis	Related to net charge.
Isoelectric Point	Isoelectric focusing	Number of charged groups.
UV-Visible Absorption Spectrum	Absorption spectroscopy	Tyrosine-Tryptophan environments, presence of absorbing ligands or impurities.[a]
Fluorescence Spectrum	Fluorescence spectroscopy	Tryptophan environment, presence of fluorescent ligands or impurities.[b]
CD Spectrum	Circular Dichroism	Secondary structure, α-helix, β-sheet, . . .
Subunits	SDS-gel electrophoresis	Quaternary structure components & their molecular weight.
Subunits & —S—S—	Gel electrophoresis in mercaptoethanol	Tertiary structure — breaks —S—S— bonds.
Solubility — Salting Out	Salting out fractionation	Solubility under different solvent conditions, pH, ionic strength, hydrophobic interactions.

[a] For example, heme in myoglobin & hemoglobin;
[b] For example, bilirubin in albumin

biochemistry texts, in particular Freifelder [10] and Cantor and Schimmel [11], are essential to the understanding of protein structure, function, and interfacial behavior. Bull's *Physical Biochemistry* is still an outstanding introduction to the subject [12]. Walton's books on biopolymers are also highly recommended [13,14], as is Tanford's book [15].

Table 1 lists those characteristics of a protein which are normally considered essential for a minimal characterization. Most of the information desired is available for practically all proteins which may be of interest to the biomaterials investigator. Generally, all of this information is available in the literature, and the investigator need only to check for purity, homogeneity, and perhaps activity of the protein preparation.

Table 2. Additional protein information desired for adsorption studies

Quantity	Method	Information/Significance
Molecular Vibrations	Raman and infrared spectroscopy	Secondary & tertiary structure.
Thermal Denaturation	Raman, CD, or fluorescence	Conformation (secondary/tertiary) & structure as f (temperature).
Monolayer Properties	Langmuir trough or pendant drop	Air/water interfacial properties, π-A curves, can deduce closepacked monolayer dimensions.
Binding Properties	Equilibrium dialysis for lower molecular weight ligands — affinity chromatography for others	Ligand binding characteristics, such as Ca^{++} binding to fibrinogen, and prothrombin, fatty acid and bilirubin binding to albumin, heparin binding to antithrombin III and fibronectin.
Amino Acid Sequence	Usually cleavage fragments plus sequencing	Sequence important in structure and function.
3-D Structure	X-ray diffraction	Complete stereo-chemistry of tertiary or quaternary structure.
Enzymatic or Specific Biochemical Activity	Substrate turnover, antibody binding, ligand binding	Is protein (enzyme) functional? Does it bind its specific ligands?

For a complete adsorption analysis, the amino acid sequence and secondary, tertiary, and quaternary structure of the protein should be known (Table 2). This is usually only available if a complete X-ray diffraction analysis has been done. Some information on ligand-binding characteristics of the protein should be available: Does is bind low molecular weight solutes? Does it have specific marcromolecular binding characteristics, etc.? For example, the physical properties of albumin are different depending on its fatty acid and bilirubin composition. The properties of hemoglobin are different if its heme ligand is in the deoxy or oxy form or in the met (Fe^{+3}) form [9]. The activity of many proteins is enhanced or potentiated when bound to appropriate ligands. For example, heparin binding to antithrombin III greatly increases the binding of thrombin. All such information is useful in predicting and interpreting protein interactions with surfaces.

The general solubility characteristics of the protein, including its behavior in different pH and ionic strength environments, its behavior in urea solutions, its behavior

Table 3. Major Non-Covalent Forces and Interactions Important in the Organization and Stabilization of Protein Structure in Aqueous Solutions [8,16,17]

Interaction	Description	Directional?	Effect of Increasing		Binding energy (kcal/mol)	Free energy change $H_2O \rightarrow EtOH$ (kcal/mol)
			Temperature	Ionic Strength or Dielectric Constant		
Ionic	Coulombic interaction due to opposite charges.	None — spherical symmetry	Decreases	Decreases	—5 $-COO^{-}...N^{+}H_3-$	—1
H-bonding	A donor-acceptor effect involving the H-atom of the donor — essentaily an ionic interaction. Water participates in H-bonds.	Yes, linear bond lengths ~3 Å; bond energies ~3 to 6 kcal/mol	Decreases	Decreases	—3 —NH...O=C—	
Charge-Transfer (Donor-Acceptor)	In water this is mainly due to π — π interactions. Trp interaction strongest as it is strong electron donor.	Yes — analagous to H-bonding	Decreases	Increases		
Hydrophobic or Entropic	Water cannot interact with apolar groups via H-bonding, resulting in water adjacent to such apolar groups being more organized or more ordered. If such apolar groups come together, some ordered water is released, increasing entropy — hence apolar groups tend to "interact." Hydrophobic interactions are particularly strong in aqueous solutions.	None	Increases	Increases; effect is ion-specific		—2.4 (phe)

in solutions containing small amounts of methanol, ethanol, or glycerol, and related information on solution characteristics are all important in interpreting and predicting interactions at interfaces.

The fundamental classes of forces and interactions important in protein structure and function are summarized in Table 3. The four general classes are: (1) ionic or electrostatic interaction, due to the attraction or repulsion of two or more groups carrying a net charge; (2) hydrophobic interaction, a largely entropically driven process due primarily to water structure effects adjacent to hydrophobic interfaces; (3) hydrogen bonding, a unique type of dipole/dipole interaction which results in interaction energies comparable to very weak covalent bonds; (4) and other interactions, primarily those of a charge-transfer or partial electron donor/acceptor type, which in aqueous solutions are often dominated by $\pi - \pi$ interactions. It is important to point out that all of these interactions take place in an aqueous medium, and that the hydrogen bonding and dielectric properties of water play a very major role (see Ref. [8,16,17]) for details).

There has been considerable effort on the prediction of secondary and tertiary structures of protein from the amino acid sequence using computeraided minimal potential energy calculations [8]). The question as to how a primary amino acid sequence begins to produce secondary and super-secondary structures and fold into its equilibrium tertiary structure and functional domains is a very active field of structural biochemistry. A related problem is the mechanism by which a protein unfolds or denatures [18–20]) which is of fundamental interest in the protein adsorption process.

A polypeptide chain in water has a specific pattern of polar and non-polar groups. There is a tendency to minimize the surface area between non-polar groups and the aqueous phase. This can be treated in terms of a surface-free energy argument or in terms of the classical hydrophobic interaction argument. The non-polar groups tend to be excluded from the surface of the polypeptide globule with the highly polar and charged amino acids on the surface interacting strongly with water. There are some regions on the surface which are non-polar, and there are some regions in the interior which contain polar groups, either through hydrogen bonding or salt bridge associations. About 90% of all internal polar groups form hydrogen bonds. The solvent entropy (hydrophobic interaction) effect contributes significantly to overall protein structural stability. The intramolecular interactions are highly efficient and effective.

Disulfide bonds confer additional stability. A frequently encountered structural component is the sequence —Cys—Cys— with both residues forming disulfide bonds with other cystines. This is a useful architectural unit and forms the basis for linking three different chain segments in close proximity. This structure is found in serum albumin.

The interior of the protein has a very high packing density, comparable to that in the protein crystal. The average packing density of a protein is about 0.75, which is incredibly high when one considers that a close-packed structure of spheres has a packing density of 0.74. Thus the packing density of proteins in terms of volume utilization is even more efficient than most close packed metals [8]). One often hears the criticism that the structure of a protein in the crystal must be very different than in the solution. This is not the case, as the solution packing of a protein is essentially identical to that in the crystal. It is important to note that active centers in reactive sites of proteins tend to be much more loosely packed than the rest of the molecule, thereby providing the increased flexibility required for chemical reactions and recognition processes.

Careful study of the structure of available proteins suggests regions of amino acid homology and structural homology in very different protein molecules. This has led to the concept of the structural and functional domain. One can consider large globular proteins as being made up of a set of functional domains, a more important structural concept than the chain composition of the protein. Active sites are often located at the interface between two structural domains. One can, therefore, consider proteins as being modular structures with the structural or functional domains as the modules. A good example is the immunoglobulin molecule shown in Fig. 1 (an IgG).

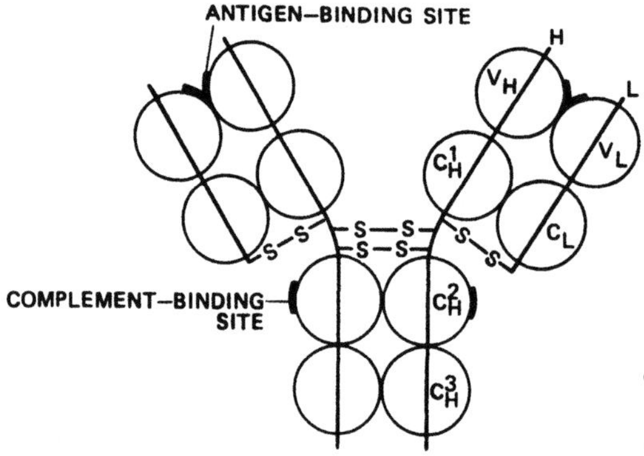

Fig. 1. Structure of an IgG molecule showing the light (L) and heavy (H) chains, the antigen-binding sites (in the Fab regions), the complement-binding site (in the Fc region), and the constant (C) and variable (V) amino acid regions (from Ref. [8])

There are both functional and structural domains. A functional domain may consist of one or more structural domains. A functional domain is a functionally autonomous region of the molecule. In the case of the IgG, the Fab portion is the functional antigen recognition domain while the Fc region is an effector domain. Structural domains are geometrically separate.

Knowing the binding characteristics of the various domains in the molecule allows the investigator to begin to deduce certain types of adsorption behavior. For example, it is known that human plasma fibronectin contains several heparin binding sites, with pK's in the 8–9 range and with a high, positive charge density. It would be reasonable to suspect that when fibronectin interacts with negatively-charged surfaces it may do so through such binding sites, just as it does with heparin.

The large globular proteins of blood plasma are monomers — that is, they consist of only one chain or a set of covalently linked chains. An oligomer is an aggregate of monomers whereby the aggregate is held together by non-covalent bonds. Most intracellular proteins are oligomers; proteins in blood plasma are large monomers, generally consisting of several funtional domains.

A structural domain can be defined as a region of locally high electron density separated by regions of low electron density or clefts in the electron density map. Amino acid residues that are far apart along the chain tend to be far apart in the three-dimensional structure. The concept of neighborhood correlation suggests, perhaps based on kinetic folding considerations, that those amino acids which are close together in a sequence tend to stay close together in the three-dimensional structure. Such neighborhood correlation is probably a consequence of the chain-folding process.

The domains are probably those pieces of a chain which fold independently. Most globular proteins have domains of 100–150 residues and are roughly 25 Å in diameter.

Five different classes of structural domains have been identified, based on the secondary structural makeup [8]. Structural symmetry is often evident, particularly in aggregate or oligomeric proteins. The more contacts, the more stable the interaction.

Depending on the fit, the number and strength of contacts, and other factors, a monomer-oligomer equilibrium will exist. For example, in the case of Concanavalin A the tetramer is in equilibrium with dimer and monomer under normal conditions. In the case of insulin the monomer, dimer, and hexamer are all in equilibrium. In hemoglobin the dimer and tetramer are in equilibrium. In the case of the adsorption of such proteins, one must not only know what is the aggregation state of the protein in solution, but must be able to deduce the adsorbed state. Do the molecules adsorb as dimers [21]? Do they adsorb as dimers and then associate as tetramers or hexamers on the surface, etc.?

Some of the free-energy considerations used to probe protein oligimerization and aggregation are also applicable to adsorption. Protein-protein interactions usually involve surface complimentarity, which can provide significant specificity. The specificity of a surface is determined by its shape, its pattern of hydrogen-bond donors and acceptors, its pattern of charges, and even its pattern of hydrophobic patches. Based on immuno globulin studies, a 100 $Å^2$ surface can form on the order of a thousand patterns showing different binding specificities [8].

An important question is how the linear, amino acid sequence spontaneously folds into the three-dimensional structure of the native form. There has been considerable work on the folding process, as well as on the unfolding or denaturation process. It is possible to take a protein in its native form, and by subtly changing its microenvironment (pH, ionic strength, temperature, addition of denaturants such as urea, addition of other ions, etc.), induce the protein to undergo a conformational change. The energetics of that change can be measured. In a number of such studies, summarized briefly in Table 4, one can conclude that the native state is only marginally stable and is in equilibrium with a variety of other states under normal conditions. One can write the equilibrium equation N $\leftrightarrows$ D. The total free energy change in going from the

Table 4. Estimates of the stability of globular proteins under physiological conditions*

Protein	Denaturant	ΔG_D^{H2O} (kcal/mol)
Lysozyme	GdnHCl or Urea	9.5
	T-pH	10.0
Ribonuclease	GdnHCl or Urea	8.5
	T-pH	6.0
Myoglobin	pH	14.0
α-Chymotrypsin	T-pH	14.0
Immunoglobulin light chain (Wes) domain	GdnHCl	5.5

* (from Ref. [20], p. 38)

native (N) state to one of the available denatured (D) states is only in the vicinity of 10 kcal/mole for an average globular protein of 150 residues [20]. This is an energy corresponding to only two to five hydrogen bonds per molecule! It is clear that protein conformations, therefore, are very fragile and truly only marginally stable. For example, the burying of just one surface tryptophan residue provides an energy of 3.4 kcal/mole. Thus, when a protein finds itself in the vicinity of an interface, its microenvironment is altered, and clearly its equilibrium will be altered, probably altering the protein conformation in the adsorbed state.

In vitro folding and denaturation studies have established that the process occurs in the time span of 0.1 to 100 seconds. It is commonly assumed that there exist folding nuclei — that is local regions of alpha helices which are stable and which function as nuclei for additional folding or organization [8]. This provides a domain, therefore, with strong local neighborhood correlation. It is expected that there are various folding pathways and intermediate structures which are used during the folding process. Some of the theoretical methods which are now being applied to the simulation of folding processes may, in modified form, have application to the simulation of the adsorption process.

A protein and its associated water molecules constitute an extremely small thermodynamic system subjected to the continual Brownian motion or buffeting of the solvent molecules [22]. Such transient changes result in fluctuations in the local thermodynamic properties of the system. A protein with a molecular weight of 25000 experiences internal energy fluctuations of about 40 kcal/mole [22]. This is much larger than the energy changes involved in ligand-induced conformational transitions. These thermal fluctuations can result in volume changes which provide transitory cavities or channels in the protein. These dynamic fluctuations are a consequence of the thermodynamic uncertainty of small systems. They exist within a single macromolecule and are not correlated with similar fluctuations in other molecules. It is clear from these thermodynamic fluctuations that conformational mobility must be expected. It is now known from NMR studies that even such large and rigid structures as the planar aromatic amino acid side-chains can flip in and out of the protein surface [23].

Proteins are therefore dynamic, flexible objects whose physical and chemical properties are dominated, not only by their conformation, but also by the continual changes in conformation which are a consequence of their microscopic size [22]. The marginal stability of most protein conformations suggests that processes at one point in the protein might well have an effect on a portion of the molecule far removed.

The process of ligand binding to a protein may also involve conformational change. The process might be viewed as follows [22]:

Freely diffusing ligand approaches the protein. Assume that the ligand binding site is not a rigid structural feature, but involves transitory conformations. If the protein is in a transitory "open" conformation, the ligand may diffuse in and bind. The binding interactions may now stabilize a conformational state which is optimum for ligand-protein interaction, a state which previously, and in the absence of ligands, would be unstable.

One can envision a conformational fluctuation spectrum with the native state being a mean conformation. Since that mean conformation is only marginally stable, it is clear that any local microenvironmental change, such as a solid surface, may well stabilize other conformations.

In order to begin to understand and appreciate how a protein may interact with a surface, it is important to be able to "see" the protein in three dimensions. This usually requires the use of models which, in the case of proteins, are terribly cumbersome and expensive. In many institutions molecular models have been replaced by three-dimensional molecular computer graphics. There are some 70 major molecular graphics installations throughout the world which have the capabilities of imaging large macromolecular structures in three dimensions.

Ideally, one should be able to predict a three-dimensional structure from the amino acid sequence. There has been considerable activity along these lines. Current methods can predict the structural class of a protein or domain. Although about two-thirds of all residues in a polypeptide chain can be assigned to the correct secondary structure, the three-dimensional structure cannot be predicted.

If one independently knows the three-dimensional structure, for example, from X-ray crystallography, i.e., all of the atoms are identified and their atomic coordinates known, then this information can be fed into a computer and the three-dimensional structure displayed. Such coordinates and related information are available for over a hundred proteins from the Protein Data Bank (Brookhaven National Laboratory, Brookhaven, New York). The structure can be displayed as a wire or stick model or as a space-filled model. With space-filled models, one sees only the outer surface of the protein. Using one of a number of algorithms for defining a protein surface [27] and by color coding the different amino acid residues or atom types present on the surface, one can quickly "see" which regions of the molecule are hydrophobic, positively or negatively charged, non-charged but polar, and other features.

Feldmann and others have developed a series of teaching tools for macromolecular structure using color-coded molecular graphics-derived images [24-28]. Connolly and others have improved upon this approach by clearly showing which portions of the protein surface are indeed accessible to a water molecule, as opposed to those portions which are inaccessible, such as in clefts, etc [27]. Images based on the Connolly methods directly show those regions of the protein surface which can be expected to interact with other molecules. It provides immediate, comprehensible information about steric complimentarity.

Another major advantage of molecular graphics is that it allows real time manipulation of several interacting molecules while quantitatively monitoring the stereochemistry and even the interaction energies. These methods are only beginning to be used for studies of interactions between molecules and only in the most preliminary way for the simulation of protein adsorption [29].

It is recommended that any reader seriously interested in protein adsorption obtain *Teaching Aids for Macromolecular Structures* [28], which is commercially available for about $20.00. These aids clearly show the dramatic potential of surface protein structural visualization for the development of hypotheses of protein-surface interactions.

How can all this protein information help our understanding of the adsorption process? Figure 2 is a schematic and idealized view of a single protein interacting with a single, well-characterized solid surface. Assume that the solid surface is well-characterized [30]. Presume that we know precisely the location of hydrophobic regions

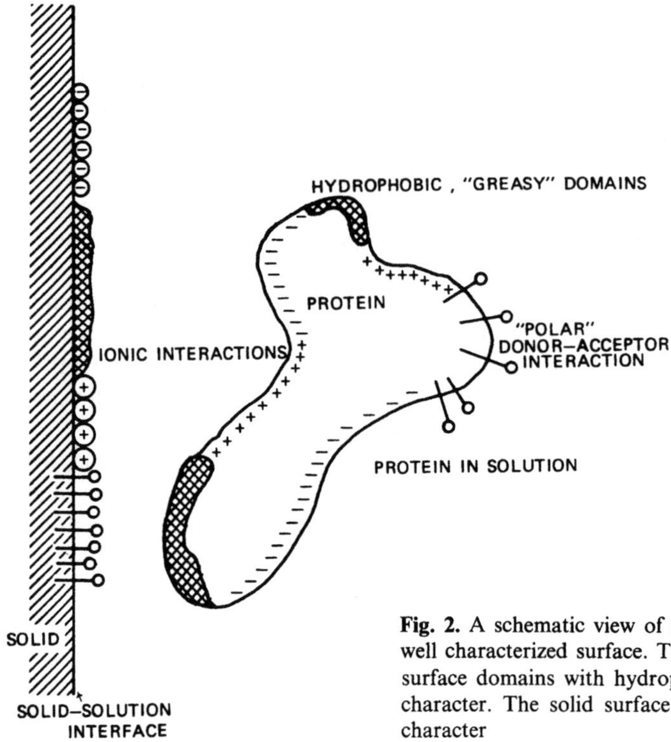

Fig. 2. A schematic view of a protein interacting with a well characterized surface. The protein has a number of surface domains with hydrophobic, charged, and polar character. The solid surface has a similar domain-like character

on the surface, the location of positively and negatively charged regions, neutral hydrophilic regions, etc.; i.e., the solid surface properties are thoroughly mapped on a submicroscopic level. We also presume that we have comparable information for the protein; we know its three-dimensional surface structure, the distribution of functional groups, and related properties. We now begin to allow the protein to approach the solid surface (by molecular graphics?) and sample the variety of potential interaction orientations. Clearly, one or more of these interaction orientations will tend to dominate on certain areas of the surface, depending on the interactions present.

It is clear, given the protein and solid surface in Fig. 2, that the protein can interact with the solid surface in a variety of different ways, depending on the particular orientation by which it approaches and the overall binding energetics. Such mental gymnastics, aided by three-dimensional graphics and at least crude estimates of interaction-free energies, should aid in the formulation of definitive hypotheses and in the explanation of data [29].

3 Mass Transport

All adsorption and desorption processes depend on transport of solute to and from the interface. There are basically four major transport mechanisms (Fig. 3):

a) diffusion;
b) thermal convection;

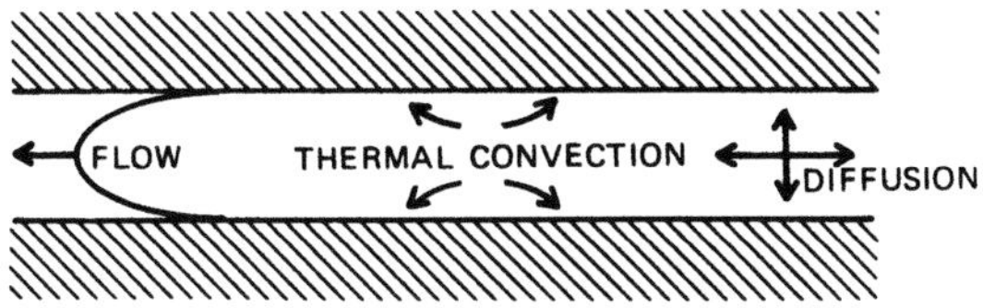

Fig. 3. The parabolic flow profile in a thin wall channel. In addition to flow, mass transport can occur by molecular diffusion and by thermal convention

c) flow (commonly called convective transport);
d) coupled transport, such as convective-diffusion processes.

Turbulent or stirred solutions may incorporate all of the processes noted.

Assume that a freshly prepared interface, such as an air/solution interface in a Langmuir trough, is optimally thermostated to minimize temperature gradients and thus minimize thermal convection. Assume there is no energy barrier to adsorption during the initial stages — every molecule which hits the interface sticks and adsorbs. All molecules near the interface will be rapidly adsorbed, resulting in a depletion of solute in the volume adjacent to the interface, called the sublayer. This concentration gradient drives diffusion from the bulk of the solution towards the depleted sublayer. Under these conditions, the rate of adsorption is equal to the rate of diffusion, dn/dt [3]:

$$\frac{dn}{dt} = C_0 \left(\frac{D}{\pi t}\right)^{1/2} \tag{1}$$

where n $=$ number of molecules;
C_0 $=$ bulk solution concentration;
D $=$ diffusion coefficient;
t $=$ time.

Thus the *rate* of adsorption is proportional to $D^{1/2}$ and $t^{-1/2}$ (Fig. 4).

Integrating gives the total number of molecules, n, adsorbed at the elapsed time (t):

$$n = 2C_0 \left(\frac{Dt}{\pi}\right)^{1/2} \tag{2}$$

The total surface concentration at time, t, is proportional to $D^{1/2}$ and $t^{1/2}$ (Fig. 4). D for proteins is in the range of 10^{-6} to 10^{-7} cm^2/s.

All fluid interfaces contain an undisturbed layer of solution adjacent to the interface. Mass transport in this boundary layer occurs only by diffusion. The thickness of the boundary layer depends on temperature, stirring, and the interface itself. It is up to 0.1 cm thick in unstirred systems and approaches 10^{-3} cm in vigorously stirred systems [3,31,32].

Once the interface is partially saturated with adsorbed solute molecules, then the rate of adsorption falls below the rate of diffusion, suggesting an energy barrier to adsorption.

Molecules at or near the interface may diffuse back into the bulk solution, particularly if the free energy of adsorption is not very high. Mass transport equations which account for back diffusion are available [3,31,32].

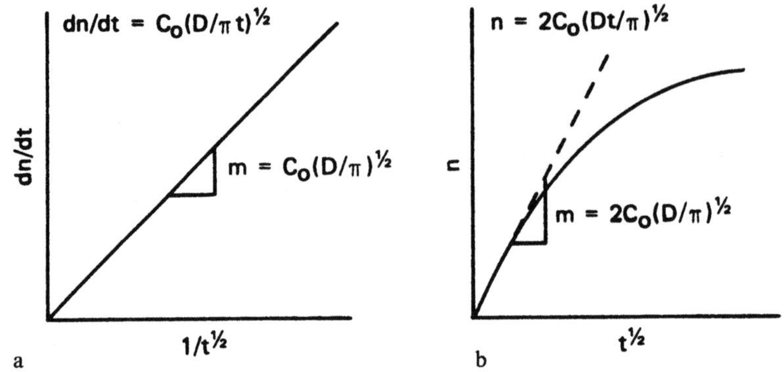

Fig. 4a and b. Plotting of diffusion-limited adsorption data (see Ref. [3]). **a.** rate of adsorption, dn/dt, is proportional to $t^{-1/2}$; **b.** the total amount adsorbed, n, in the initial stages is proportional to $t^{1/2}$

Adsorbed molecules may diffuse laterally at the interface. Although surface diffusion is well-known in classical surface chemistry [33], data on adsorbed macromolecules is sparse. Burghardt and Axelrod [34]; and Michaeli et al. [35] have both demonstrated rapid interface diffusion of adsorbed albumin.

In many situations, the system is primed with a buffer solution which is displaced by the protein solution of interest (Fig. 5a). Assuming constant, laminar established flow, the velocity (V) in a rectangular flow channel of width (w), thickness (b), and length (l), where $b \ll w$ has a characteristic parabolic profile, given by [36]

$$V = 6\bar{V}\left(\frac{y}{b} - \left(\frac{y}{b}\right)^2\right) \tag{3}$$

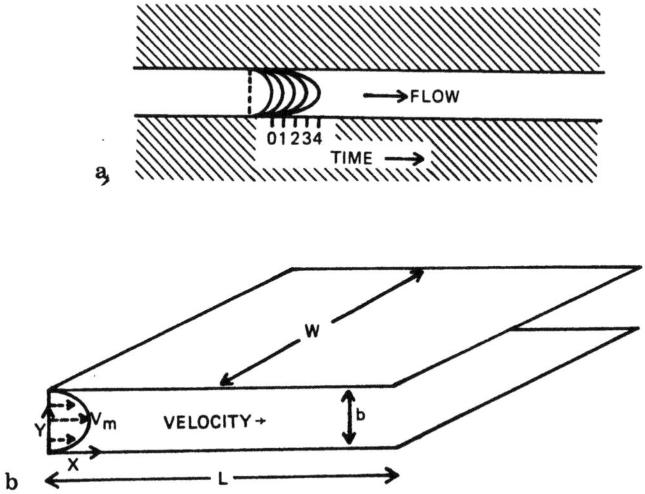

Fig. 5a. The development of the concentration profile due to a plug of protein solution entering a buffer-primed, thin plate flow channel. Note that a "bullet-shaped" concentration profile develops with time (assuming no diffusion); **b.** the geometry and coordinate system used in the convection-diffusion treatment

where $\bar{V}$ is the average velocity and y is the y coordinate (see Fig. 5b).
The volumetric flow rate, $Q = A\bar{V}$, where A is the cross sectional area of the channel, i.e., $A = b \cdot w$, therefore,

$$V = \frac{6Q}{b^2 w} y \left(1 - \frac{y}{b}\right) \tag{4}$$

It can be shown that the wall shear rate, γ, is

$$\gamma = \frac{6Q}{b^2 w} \tag{5}$$

thus

$$V = \gamma y \left(1 - \frac{y}{b}\right) \tag{6}$$

which is the equation for the parabola shown in Fig. 5.

Physically, the shear rate, γ, is the change in fluid velocity over a change in distance, dV/dy, or the velocity gradient. The units of γ are

$$\frac{\text{cm/s}}{\text{cm}} \text{ or } s^{-1}$$

The shear stress is also of interest. Shear stress (τ) for Newtonian fluids is linearly related to γ,

$$\tau = \mu\gamma \tag{7}$$

where μ is the viscosity.

Note that the velocity at the wall ($y = 0$ and $y = b$) is zero, meaning nothing moves at the wall, according to classical fluid mechanics.

Now imagine that a protein solution enters the flow chamber at time $t = 0$. The protein solution (of uniform concentration and flow velocity) begins displacing the buffer solution (Fig. 5a). Given a parabolic velocity profile, Fig. 5b shows the development of the concentration profile in the cell at various times after entrance. A "bullet-shaped" concentration profile develops. This is easily observed using a dye solution or blood. No protein reaches the surface by convective flow alone. Protein is transported to the interface by diffusion.

The flow system can be designed to produce mixing which minimizes concentration profiles. Such processes are discussed in any basic fluid mechanics or flow injection analysis text book [36,37].

A complete treatment of mass transport to interfaces requires combining convective (flow) and diffusion processes [36–38] as the molecules present in a flowing stream are transported by flow (convection) and by diffusion simultaneously. Fortunately, this mass transport problem is well-treated [38], particularly in the chemical engineering literature. Robertson [39,40] and Leonard [41] are chemical engineers who have been very productive in the area of protein adsorption — they and their students have presented and used convective diffusion treatments extensively.

For the flow direction and coordinate system given in Fig. 5b, and assuming diffusion perpendicular to the solid-liquid interface, the general convective diffusion equation in Cartesian coordinates is [38-40]:

$$\frac{\partial C}{\partial t} + V(y)\frac{\partial C}{\partial x} = D\frac{\partial^2 C}{\partial y^2} \tag{8}$$

where

$$V(y) = \gamma y\left(1 - \frac{y}{b}\right) \tag{9}$$

C = concentration,
t = time,
D = diffusion coefficient.
The origin of the coordinate system is the entrance to the flow chamber (far left of Fig. 5b). Assuming there is no adsorption at the interface, the boundary conditions are [40,42]:

at $t = 0$, $C = 0$ for all y, $x > 0$;
at $x = 0$, $C = C_0$ for all y, $t > 0$;
at $y = b/z$, $C = 0$ for $t < x/Vm$;
$C = C_0$ for $t \geqq x/Vm$;
at $y = 0$, $\partial C/\partial y = 0$ for all $x > 0$, $t > 0$.

The equation can be solved using various numerical methods.

As the protein solution moves into and through the rectangular flow channel, a parabolic concentration profile develops (Fig. 5a). Diffusion washes out the flow profile, eventually equalizing the concentration at C_0.

In the absence of diffusion, the injected material would never reach the wall due to the laminar nature of the flow. In the presence of diffusion, the concentration at the wall rises with time in an S-curve fashion. Lok generated a dimensionless equation which defines the time course of diffusion [40]:

$$\tau = t\gamma^{2/3}\frac{D^{1/3}}{l^{2/3}} \tag{10}$$

where l is the distance from the entrance to the midpoint of the flow chamber and where $\tau = 3$ to reach a plateau value (i.e. the concentration at the wall is C_0). Solving for t, we find that

$$t = \tau\frac{l^{2/3}}{\gamma^{2/3}D^{1/3}} \tag{11}$$

Using the typical values $D = 4.0 \times 10^{-7}$ cm^2/s for gamma globulin, $l = 2.5$ cm, $\gamma = 47.4$ s^{-1} for a volume flow rate of 2.0 ml/min and $\tau = 3$ to reach equilibrium, the diffusion time is 57 seconds from the time the newly injected fluid enters the cell.

The study of adsorption or desorption kinetics in the first minute or so must take full account of convection-diffusion processes. Fortunately this has been well-modeled by Lok, et al. [40] and by Watkins and Robertson [39,42].

In any adsorption situation, the adsorption process itself is involved in the boundary conditions, thus affecting the solution of the equation [39-41].

Limiting cases of the general convective-diffusion equation are often helpful. If the time dependence is ignored, i.e., $\partial C/\partial t = 0$ (for example, at low bulk protein concentration, at long times, and/or when the rate of adsorption is much greater than the transport to the surface), then we have

$$V(y)\frac{\partial C}{\partial x} = D\frac{\partial^2 C}{\partial y^2} \tag{12}$$

This simpler expression can be solved numerically with the appropriate boundary conditions. This is the so-called transport limited or Leveque case [38,41].

Another limiting case is when $V = 0$ (no flow), then we get the common diffusion expression

$$\frac{\partial C}{\partial t} = D\frac{\partial^2 C}{\partial y^2} \tag{13}$$

Techniques which allow one to monitor the boundary layer as a function of time, such as total internal reflection fluorescence (TIRF) spectroscopy [4,43], permit a quantitative evaluation of interfacial mass transport processes using, for example, fluorescently-tagged macromolecules which do not adsorb, such as fluorescein-labeled dextran [40].

Vroman [44] has recently demonstrated the great importance of volume and concentration in limiting interfacial transport and thus in influencing the adsorption process (see also Sect. 4.6).

4 Protein Adsorption Principles

4.1 Background

The classical Langmuir theory for gas adsorption can be applied to adsorption from solution, if the solution is sufficiently dilute [45,46]. The surface is considered to consist of sites of about the same area as the projected area of the solute of interest (Fig. 6).

Let $\bar{v}_s$ be the number or moles of solute molecules adsorbed per area of surface. The subscript on $\bar{v}$ refers to the case of adsorption on a *solid surface* as contrasted with $\bar{v}$ used in the case of protein-ligand equilibria [4,47-49]. Let [A] be the equilibrium solute concentration.

The classical Langmuir adsorption isotherm is then

$$\bar{v}_s = \frac{kA}{1 + kA} \tag{14}$$

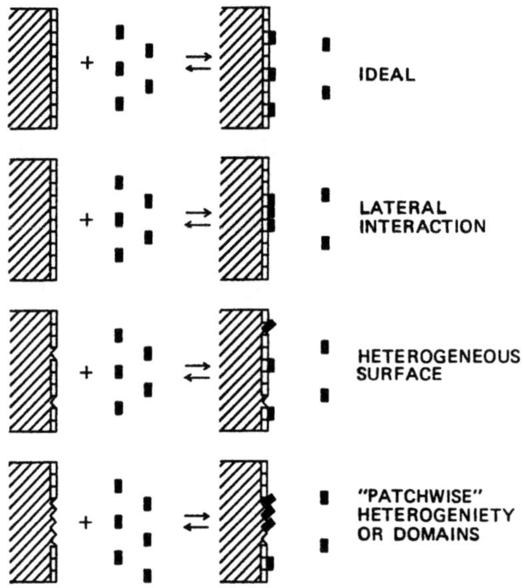

Fig. 6a–d. Schematic view of adsorption from solution onto smooth, planar surfaces where the surface "sites" are considered to have the same area as the projected area of the solute of interest. **a**. Top, the ideal (Langmuir) case; **b**. clustering of adsorbed solute due to attractive lateral interactions or positive cooperativity; **c**. heterogeneous surface, i.e., two sets of binding sites; **d**. "patchwise heterogeneity" or surface domains of different adsorptive properties

which assumes:

1. only one molecule can be adsorbed per site (commonly called the monolayer assumption);
2. only one type of site is present (homogeneous surface);
3. the adsorption of one molecule does not affect the adsorption energy of other molecules (no lateral interactions or cooperativity);
4. only one adsorbing species is present (no competitive adsorption);
5. dilute solution; and
6. reversible adsorption.

These assumptions are identical to those in the ideal theory of multiple equilibria used to analyze low molecular weight ligand binding to proteins [4,47–49]. Figure 7 shows how these assumptions can be considered in more refined treatments of adsorption from solution [45].

The case of the heterogeneous surface (Fig. 6) is treated as

$$\bar{\nu}_s = \sum_i \left(\frac{k_i A}{1 + k_i A} \right) \tag{15}$$

The case of patchwise heterogeneity versus "homogeneous" heterogeneity cannot be distinguished by adsorption data using low molecular weight solutes [45]. The case of lateral interactions can also be treated using assumed functions for the lateral interaction energy (Figs. 6 and 7) [45].

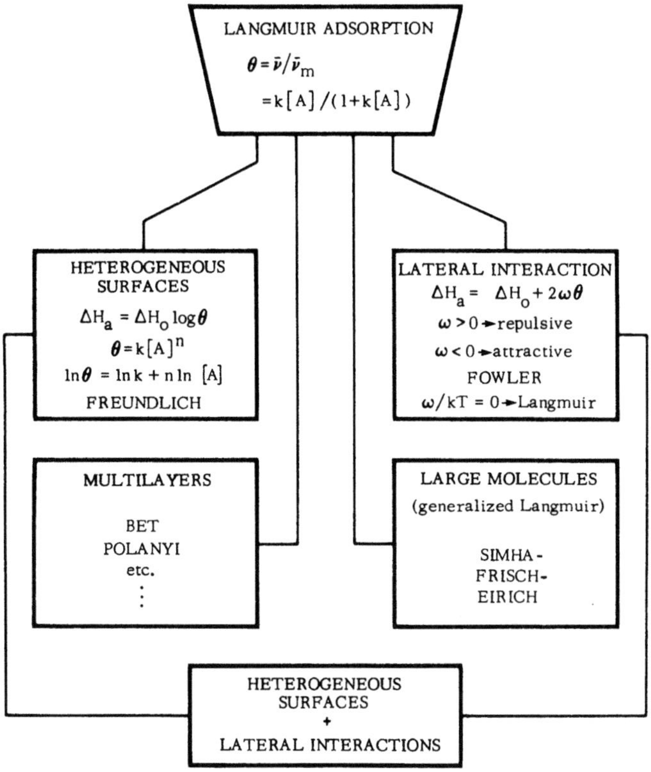

Fig. 7. The relationships between the different models of adsorption. The various assumptions involved in the classical Langmuir treatment can be considered explicitly in other, more complex models (see Ref. [45]) for further details)

The adsorption theories and protein-ligand equilibria theories are formally identical. If one simply treats the protein molecule as a heterogeneous surface (two or more classes of binding sites), interacting with identical low molecular weight ligands, the equations are identical. The assumptions involved are generally identical — for example, one ligand per site, which is the same as one adsorbed molecule/site or the monolayer assumption of Langmuir adsorption. Lateral interaction in the Langmuir treatment becomes cooperativity in the ligand equilibria treatment; heterogeneous surfaces are the same as two or more classes of binding sites; etc.

There is a class of adsorption or ligand equilibria data which is of particular importance and relevance to our discussion — the literature of protein chromatography. Proteins can be separated by ion exchange, hydrophobic, and charge-transfer processes on suitable chromatographic supports. Generally, a support is used which has very low interactions with protein (k is very low). A common one is cross-linked agarose. The support is then derivatized with the proper type and density of binding site (immobilized ligand). The ligand may be a carboxylic acid group, a quaternary ammonium group, a charge-transfer (electron donor or electron acceptor) group, or a hydrophobic group. We shall only consider the hydrophobic case as an example here. The underivatized agarose is a high water content gel which consists mainly

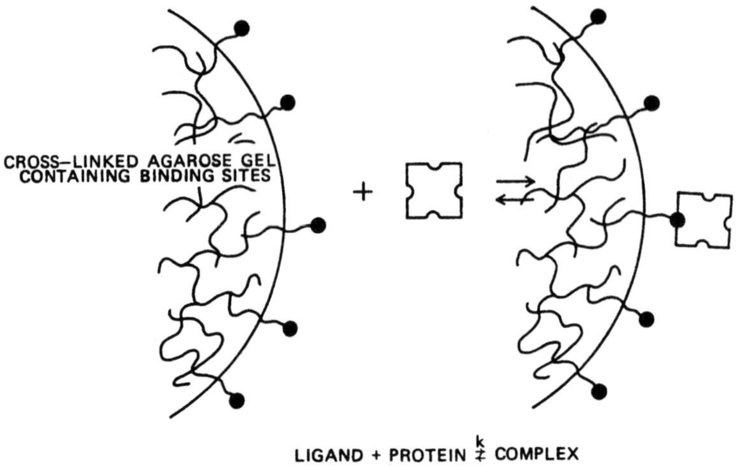

Fig. 8. Schematic of a hypothetical protein interacting with a crosslinked agarose gel containing certain bonding sites. The bonding sites on the gel are considered to be immobilized ligands

of —OH groups, which are considered to have a negligible interaction with most proteins in water.

Consider Fig. 8 where the binding equilibrium is expressed diagramatically. Clearly the binding groups on the agarose bead surface play the same role as ligand in a multiple equilibria treatment. Jennissen [50] has used this approach to study the interaction of protein with alkyl-substituted agaroses of different degrees of substitution (alkyl group densities or surface concentrations). His data [51] suggests a cooperative binding process, where about four butyl residues (binding sites) are involved in the adsorption. In the case of the methyl derivative of Sepharose[1], 6 to 7 methyl groups are required for adsorption [51].

The usual adsorption experiment consists of a fixed surface and a variable protein concentration, thus, the "ligand" concentration is constant and the protein concentration is varied. The expressions can be derived in terms of [P] rather than [A], where [P] is the equilibrium protein concentration [4]:

$$\bar{v} = \frac{\text{moles bound protein}}{\text{total moles ligand}} = \frac{[PA]}{[A_T] = [PA] + [A]} \tag{16}$$

For n identical, non-interacting sites

$$\bar{v} = \frac{nk\,[P]}{1 + k\,[P]} \tag{17}$$

Due to steric considerations, n will generally be a lower number than in the free ligand case. For the adsorption of proteins on a solid surface, the total ligand concentration is usually represented as the actual surface area of the adsorbent. In studies utiliz-

[1] Sepharose is the commercial cross-linked agarose produced by Pharmacia, Inc.

ing cross-linked gel chromatographic adsorbents, ligand concentration is often given as moles/ml packed gel. Units of [k] are reciprocal concentration units.

Jennissen's data [50, 51] shows that in terms of ligands bound (for example, butyl groups), the process is positively cooperative. Binding of protein to the first butyl group increases the probability of interaction with a second butyl group, etc. However, in terms of protein binding or protein adsorption, binding of one protein to the butyl groups decreases the probability that a second protein can interact, or the system exhibits negative cooperativity. Figure 9 summarizes these concepts in simple, diagrammatic form.

The adsorption or interaction of proteins with chromatographic supports discussed in the previous section made one overriding assumption: the underivatized support does not significantly interact with proteins; i.e., the underivatized agarose has such

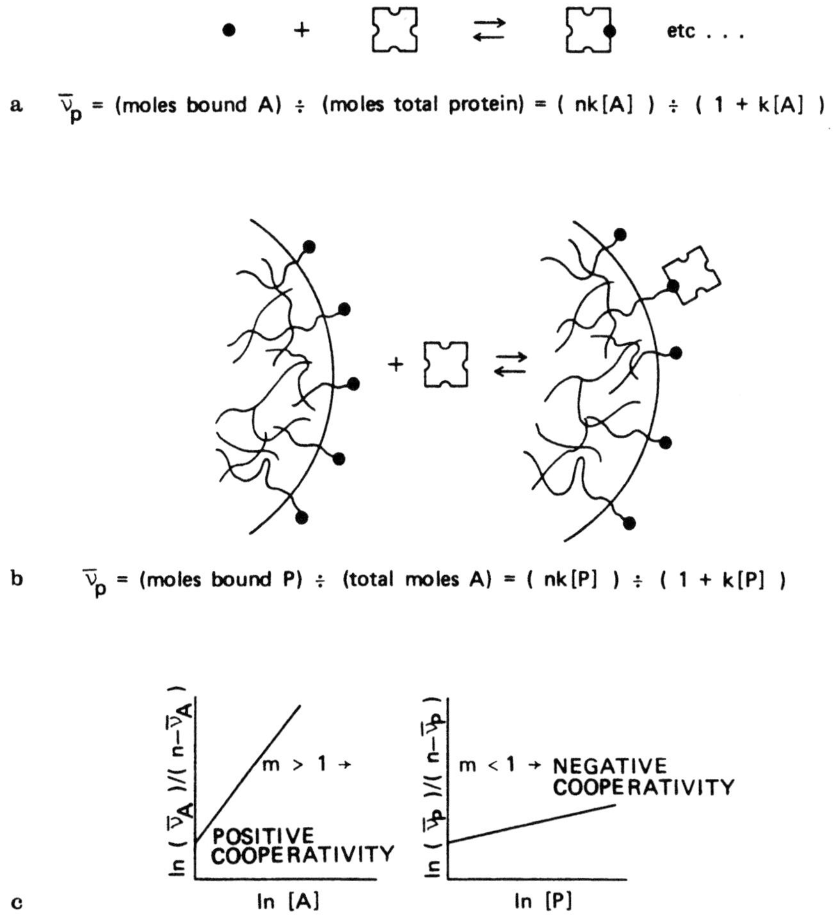

Fig. 9a. Top: Protein — soluble ligand classical binding equilibria and definition of $\bar{\upsilon}$ as moles bound ligand/moles total protein; **b.** Center: Protein adsorption — binding equilibria now treated in terms of soluble protein concentration; $\bar{\upsilon}$ now defined as moles bound protein/moles total immobilized ligand. **c.** Bottom: Hill plots for each case, demonstrating that positive cooperativity with respect to binding of *ligand* is negative cooperativity with respect to binding (adsorption) of protein

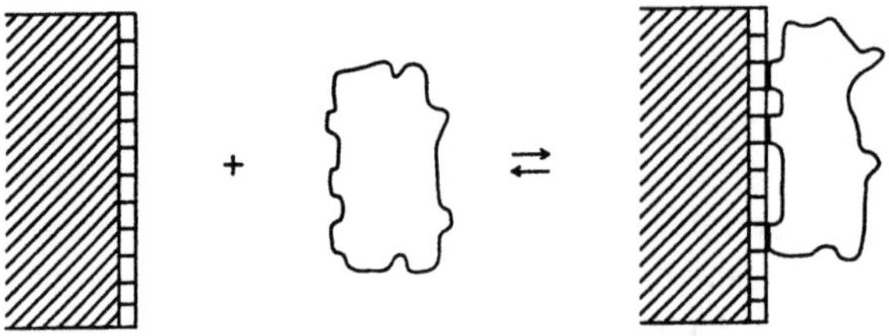

Fig. 10. Typical adsorbent surfaces can be considered to have very high binding site ("ligand") densities, resulting in multivalent interactions with adsorbed protein. If the multivalent interactions are of sufficient number and energy, the adsorptive interactions is "irreversible"

a low interaction energy per site with proteins that for all practical purposes that interaction is negligible. This is not the case with most other surfaces with which proteins come in contact. Using our earlier definition of surface site from Fig. 6, Fig. 10 schematically shows what happens when a typical surface of very high site density interacts with a protein. This is the case with hydrophobic surfaces such as, for example, polyethylene, polypropylene, polystyrene, and related biomedical polymers. One can consider every monomer unit exposed at the surface in such polymers to be a surface site interacting with the hydrophobic domains or patches on a protein by strong hydrophobic interactions. A large number of such interactions per individual protein molecule results in irreversible adsorption. This is, of course, also experienced in conventional protein adsorption chromatography, whether by hydrophobic or ionic (electrostatic) mechanisms. When the site density on the chromatographic support is too high, the protein is irreversibly bound. Sometimes this condition can be overcome by changing the elution medium. In the case of ion exchange chromatography, one would go to a higher ionic strength eluant, which would decrease the magnitude of the electrostatic interactions. Under these new ionic strength conditions, that protein would then be reversibly bound and could be eluted. The same is true of irreversible hydrophobic adsorption. If one elutes with an eluant which destroys or decreases the magnitude of the hydrophobic interaction, then what was an irreversible process is reversible in the new solution environment [52]. Most studies of protein adsorption onto solid surfaces in the literature are under irreversible or at best semi-reversible conditions.

Ideally, protein adsorption experiments should be conducted on surfaces of low binding site density so that the interaction remains reversible and can be analyzed by multiple equilibria and related models. This will then allow the estimation of the number of interaction sites and information on binding constants and interaction energies. This is not only useful for the characterization of the surface, but for the characterization of the protein itself, i.e., the accessible regions on the protein surface which are involved in the adsorption process. There are cases and conditions in which proteins can be studied on hydrophobic surfaces in a reversible fashion — for example, proteins which appear to be very weakly hydrophobic or at the bottom of the scale in hydrophobic chromatography experiments [52]. Albumin is highly hydrophobic and near the top of the hydrophobic chromatography scale and would be expected to adsorb irreversibly to most low energy surfaces. On highly hydrophilic surfaces, such as clean, wettable quartz, it is known that some proteins adsorb in a highly reversible fashion, including albumin [53]. Here, the hydrophobic interactions are probably minimal, and the major adsorptive interaction is probably ionic.

The most useful information on the interactions of proteins with surfaces will come from studies analogous to those of protein chromatography, where well-characterized and understood proteins are studied with well-characterized surfaces of known functional group type and density. The information obtained is then analyzable in such a way as to deduce interaction site densities and interaction energies. Only with such data in hand will we be able to begin to quantitatively treat and understand protein adsorption.

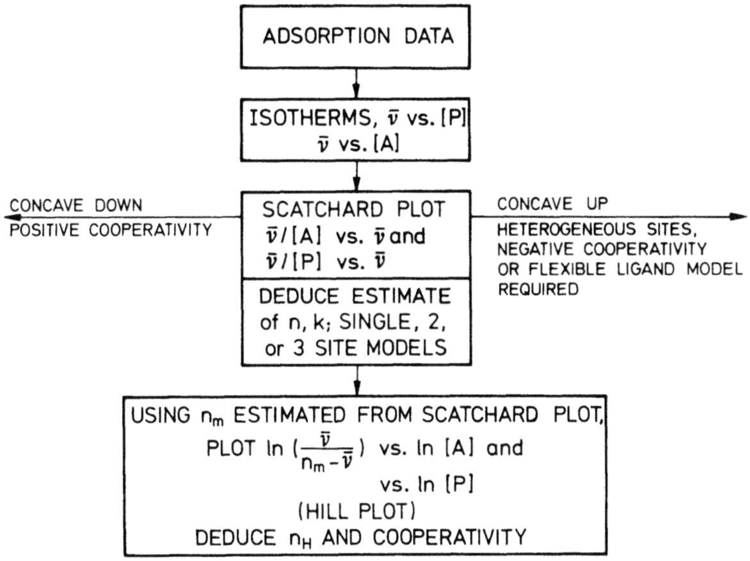

Fig. 11. Scheme for the analysis of protein adsorption data using the methods of multiple equilibria of proteins

Given the assumptions and limitations outlined in this section and elsewhere [4], conventional adsorption isotherm data can by analyzed in terms of multiple equilibria models. The general scheme is outlined in Fig. 11. First, the adsorption data is taken with sufficient care to get a wide range of points over a wide range of concentrations, preferably as both a function of protein solution concentration and immobilized ligand concentration, as outlined by Weber [54, 55]. In many cases, the ligand surface concentration will be kept constant or will be largely unknown. The adsorption isotherm is then plotted. It is then most convenient to perform a Scatchard plot [4]; if the plot is approximately linear, an estimate of n and k can be derived. If the plot has a strong concavity, then one can use the approximation techniques to derive a set of n and k for two different assumed site populations. The shape and concavity of the Scatchard plot immediately indicates several things. If it is concave-down, one can assume that this is due primarily to a positive cooperativity process. For example, the papers by Jennissen make such conclusions [50, 51]. If it is concave-up, one can assume two or more populations, i.e., heterogeneous sites. One can also assume negative cooperativity. The concavity may be due to the fact that one is using a flexible ligand, and a flexible ligand model is needed to analyze such data [56]. However, from the Scatchard plot or from the original isotherm, one can generally obtain an estimate of the upper value for n. Using this value for n, one can perform a Hill plot, i.e. log $\bar{\nu}$/n-$\bar{\nu}$ versus log [P] [4,48,49]. From the slope of the Hill plot one can deduce some measure of cooperativity.

For many proteins and many surfaces, the adsorption will be essentially irreversible, which will result in a Scatchard plot with a shape which is not easily analyzed. The models and methods presented here are not adequate for treating such data. It is important under such conditions to redo the adsorption experiment, preferably on a modified surface of lower binding site density or energy. Figure 12 outlines in schematic form what one might expect if this is done, although no such data are at present available in the literature.

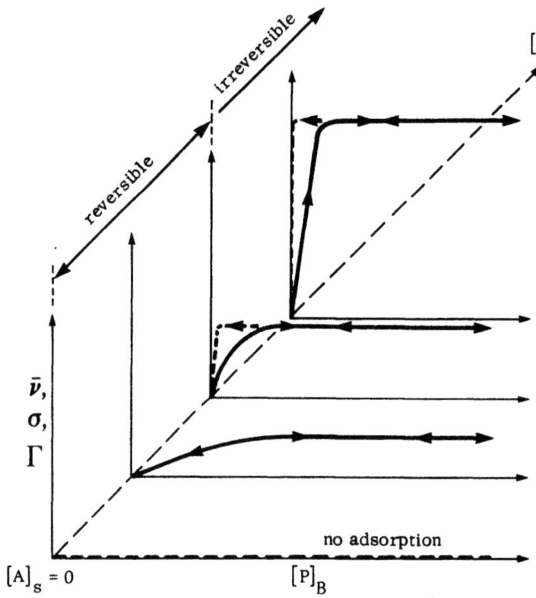

Fig. 12. Hypothetical 3-D plot of protein adsorption isotherm ($\bar{\upsilon}$ plotted against $[P]_B$) as a function of surface ligand concentration, $[A]_s$. Note that the system is reversible only up to a critical $[A]_s$ and then behaves irreversibly for higher ligand surface concentrations. The right arrows (→) denote adsorption; the left-facing ones (←), desorption

Assume that we have a surface — for example, a substituted agarose analogous to the chromatography supports discussed earlier — in which the ligand surface density is allowed to vary. Further assume that each of these different supports of different ligand group density is studied with respect to different protein solution concentrations, and the amount adsorbed is determined. One then has two sets of isotherms, basically $\bar{\nu}$ vs. [P], as a function of ligand group density or $\bar{\nu}$ vs. [A], i.e., ligand group density or concentration, as a function of [P]. These can be plotted in a three-dimensional form as indicated in Fig. 11 and analyzed by the methods described herein. This has been described briefly by Weber [55]. Note from Fig. 12 that we are assuming that as the ligand group density [A] increases, the adsorption isotherm becomes irreversible, which is no longer within the domain of analysis of the models and methods presented here. However, the adsorption behavior can be shifted to a reversible domain if the solution conditions are changed as discussed earlier.

There are some inherent limitations and assumptions in the multiple equilibria treatment. One is the fact that the treatment was devised primarily for ion interactions with proteins, that is small, spherically, symmetric rigid solutes. There are models available for non-rigid solutes [55-59], which are more appropriate to fatty acid binding to proteins, particularly through the hydrophobic alkyl chains. Such models would be more appropriate to the study of protein interactions with highly mobile surfaces, such as those of elastomers and low glass transition materials, and probably even the substituted agaroses. Finally, both the rigid solute and the soft solute models [55-59] assume that the protein is conformationally rigid, although the rigid model has been treated in terms of allosteric phenomena or conformational coupling of the binding process [60]. In most cases of particular interest, the protein cannot be assumed to be conformationally rigid, and further methods, models and means of obtaining data will be required in order to interpret and understand such processes. The allosteric treatments of conventional rigid ligand-protein equilibria [60], coupled with the treatments of mobile ligand-rigid protein equilibria [56,57] should allow such problems to be approached. The methodologies and discussion in this section, therefore, are but a first step in the analysis of protein interactions at solid-liquid interfaces.

4.2 Thermodynamics

What is the driving force for protein adsorption? Is the adsorption driven by overall energetic (enthalpic) interactions or does the entropic contribution prevail? Do both entropic and enthalpic contributions play a major part in the adsorption process, the extent of each depending on the particular protein and surface in question? An illuminating thermodynamic analysis given by Norde and Lyklema [62,66] for the adsorption of two different globular proteins (human serum albumin, HSA, and bovine pancreatic ribonuclease, RNase) on polystyrene latices will be presented. We believe this analysis has general validity.

Two parts of the adsorption isotherm can be distinguished: the initial part and the plateau region. At very low protein solution concentration, the adsorbed amounts are so low that lateral interaction between protein molecules can be neglected. This part of the isotherm (or its initial slope) reflect only the interaction between protein and adsorbent. In the case of HSA adsorption on negatively-charged polystyrene latices at pH = i.e.p., the adsorption was found to be independent of temperature, T, while at pH greater or less than the i.e.p., adsorption increases with increasing temperature [68]. The positive values of $(\partial \Gamma / \partial T)$ suggest that the process is enthalpically unfavorable and is entropically driven. The more the pH differs from the i.e.p., the larger $(\partial \Gamma / \partial T)$ becomes, implying that unfolding of HSA on the surface is a process which is not favored enthalpically.

At the plateau region direct calorimetric measurements of $\Delta_{ads}H$ (adsorption enthalpy) showed that for a number of protein-surface combinations there is a part of the pH range where $\Delta_{ads}H > O$. Since $\Delta_{ads}G < O$ for the process to occur, $T\Delta_{ads}S$

> O. The positive sign of the entropy change indicates that there are conditions where the process of adsorption is entropically favorable, if not driven solely by positive $\Delta_{ads}S$. For rigid proteins like RNase, unfolding is opposed by strong intramolecular enthalpic forces. If the RNase molecule cannot gain entropy by adsorption or cannot find other enthalpy sources, it can not and does not adsorb. This was found in the case of interactions between positive-charged haematite and RNase; no adsorption took place due to electrostatic repulsion [68]. Equal signs of the surface charges and the protein charge does not necessarily imply an absence of adsorption, however. It was found in a number of cases that negatively-charged HSA adsorbs on negatively-charged surfaces with its negative groups oriented toward the surface. This was due to counterions incorporated in the adsorbed protein layer between the surface and protein molecule providing an electrostatic bridging mechanism [69]. It is an important part of the adsorption process since it contributes both to $\Delta_{ads}H$ and $\Delta_{ads}S$.

Following are the contributions to enthalpy and entropy of protein adsorption that have to be considered, according to Norde and Lyklema [61,62]:

1. Hydrophobic dehydration results from bonding of the protein's hydrophobic patches to the hydrophobic regions on the adsorbent. The enthalpic part of this interaction is small; the entropy change is positive. Hydrophobic dehydration is relatively unimportant for hydrophilic surfaces and/or rigid hydrophilic proteins.
2. Overall electrostatic interactions depend on the surface charge and protein charge, both of which are usually functions of pH and solution ionic content. It is a more decisive parameter for rigid protein because of the ion-incorporation and possible conformational changes of flexible proteins.
3. Protein conformational changes contribute positively both to the enthalpy and entropy of adsorption. Such contributions are pH-dependent.
4. Ion incorporation has two important aspects: a) due to the transport of ions from solution to the adsorbed protein layer, contributions to enthalpy and entropy are usually negative; b) in the case of charge redistribution in the electrical double layer, the ion contribution to the enthalpic forces depends on the protein-surface charge difference, pH, and solution ionic content. Its entropic contribution is positive.
5. Overall, Van der Waals interactions are relatively insignificant due to the similar Hamaker constant of proteins and water.
6. Specific binding may also play an important role and its contributions to both enthalpy and entropy of the adsorption depend on the specific protein-surface combination.

Apparently, no single factor can be used to predict the process of adsorption; there are always several different properties of protein and adsorbent that determine the protein-surface interaction. As a summary, the following general guidelines can be given:

1. The affinity of a negatively-charged surface for a given protein increases if the surface is more hydrophobic or/and has smaller electrokinetic potential.
2. The affinity of protein towards a given negatively-charged surface increases the more hydrophobic the protein is and/or the smaller the amount of negatively-charged parts of protein occupy the inner region of the adsorbed protein layer.

3. Adsorption is enhanced by higher concentration and valency and smaller chaotropicity of the indifferent ions.
4. The adsorption process is often entropically driven with the gain in entropy arising from dehydration of the adsorbent surface and structural rearrangements inside the protein molecule (the state of hydration and field overlap changes inside the adsorbed protein included).

4.3 Kinetics

The kinetics of single protein adsorption have been discussed and modeled by several groups [70-73]. The models are summarized in Figs. 13a–f. These models assume the process is surface reaction limited, i.e. protein transport to the interface is not rate limiting. The nomenclature is:

C_0: bulk solution concentration [g/l]
$C_s(t)$: adsorbed surface concentration at time t [kg/m²]
$(C_s)_\infty$: plateau surface concentration at long t [kg/m²]
k_j: rate constants [min⁻¹ or 1/g min]
n_i: number of molecules in state i per unit area of surface
a_i: surface area fraction occupied by molecules in state i
n_s: total number of adsorbed molecules per unit area of surface

Figure 13a shows basic Langmuir reversible adsorption, discussed earlier.

The Lundström model [70] is given in Fig. 13b. He assumes that protein adsorbs with a rate constant k_a into State 1. Upon adsorption, some of the adsorbed proteins in State 1 (native) may conformationally change (via rate constant k_r) to State 2 (denatured). Letting n_1 and n_2 be the number of molecules per unit area in States 1 and 2, and a_1 and a_2 be the area fractions occupied in each state, he says (noting that the unoccupied area fraction $= (1 - a_1n_1 - a_2n_2)$ that [70]:

$$\frac{dn_1}{dt} = (k_aC_0 - k_rn_1)(1 - a_1n_1 - a_2n_2) \tag{18}$$

The build-up of proteins in State 2 is given by

$$\frac{dn_2}{dt} = k_rn_1(1 - a_1n_1 - a_2n_2) \tag{19}$$

Lundström shows that [70]:

$$n_2 = \frac{k_aC_0}{k_r}\ln\left(\frac{k_aC_0}{k_aC_0 - k_rn_1}\right) - n_1 \tag{20}$$

He goes on to determine relations for a_2/a_1 at steady state (assuming monolayer coverage) and for $n_1 + n_2$, given experimentally by radioisotope or ellipsometry studies.

The Lundström model allowing for desorption is given in Fig. 13c, although it has not been treated in the literature.

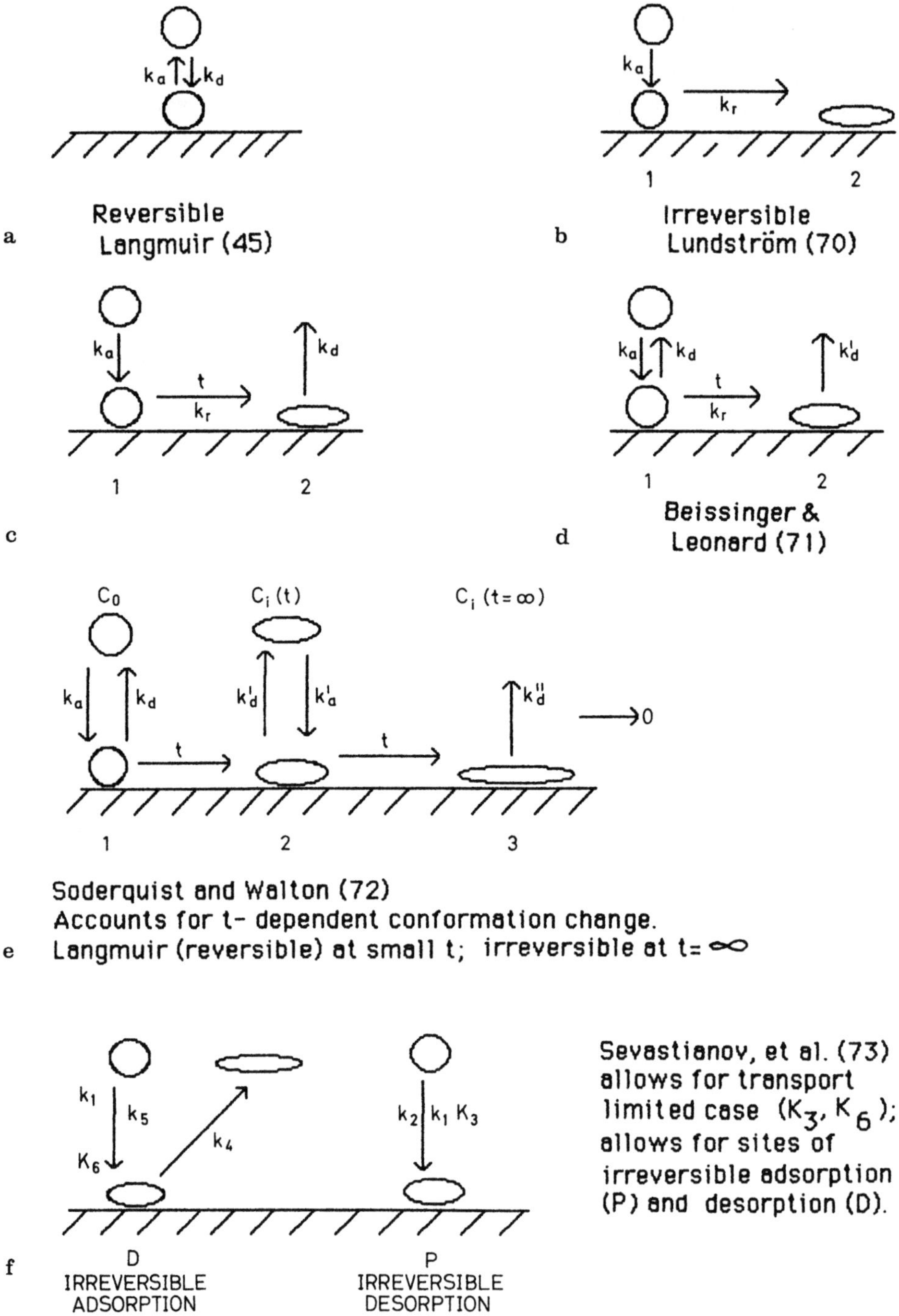

Fig. 13. Most of the kinetic models which might be applicable to protein adsorption (see Refs. [70–73]); k is rate constant, subscript a and d are adsorption and desorption respectively, 1 and 2 are adsorption states — usually native and denatured

Beissinger's and Leonard's model (Fig. 13d) accounts for desorption of both native and denatured adsorbed species (States 1 and 2, respectively). They used the classical Langmuir-Hinshelwood model for catalytic reactions (the surface is the catalyst for conformational change or denaturation of the adsorbed protein), which assumes equilibrium at a steady state between adsorbed and solution molecules. They show:

$$\frac{da_1}{dt} = C_0k_a(1 - a_1 - a_2) - a_1k_r(1 - a_1 - a_2) - a_1k_d \tag{21}$$

$$\frac{da_2}{dt} = a_1k_r(1 - a_1 - a_2) - k_d'a_2 \tag{22}$$

$$n_s = n_1 + n_2 \tag{23}$$

The equations were best fit to albumin on quartz adsorption data by a nonlinear regression method to obtain values of k_a, k_d, k_d', k_r, and n_s. They then extended the model to consider the competitive adsorption of two proteins [71].

Soderquist and Walton [72] added time dependence of the surface reaction to the model. They also allowed for readsorption of desorbed material (Fig. 13e). This model in principle takes into account the time dependence of the conformational change of adsorbed protein. They consider three distinct processes or states:

1. Rapid and reversible adsorption reaching a pseudo-equilibrium during the first minute;
2. Surface-induced change in conformation which optimizes protein-surface interaction and decreases the probability for desorption;
3. At $t \to \infty$, adsorbed material is fully denatured and if desorbed can not readsorb; desorption from State 3 is irreversible.

Process 1: Rate of adsorption, $\frac{dn_1}{dt}$, is proportional to solution concentration, C_0, but desorbed material readsorbs at a different rate.

$$R_{ads} = (k_aC_0(t_0) + k_a'C_1(t_1) + \ldots)\,A = A\,\Sigma\,k_a^{(i)}C_i(t_i) \tag{24}$$

where i refers to solution species which had been on surface for t_i; A is available surface area; k_a^i is the rate constant for species i.

Process 2 is difficult to model.

Process 3: The desorption rate is much slower, particularly after long contact times. They assume that fully denatured protein does *not* desorb.

$$R_{des} = \Sigma\,k_d^i(C_s)_i\,(t_i) \tag{25}$$

where $(C_s)_i$ is the adsorbed concentration which has resided for a period t_i on the surface. If we consider material incubated for a definite time, t, and all desorbed material is continuously removed, then

$$R_{des} = k_d C_s(t) \tag{26}$$

which can be written in the form

$$R_{des} = k_d C_s(1 + k_r t^{n-1}) \tag{27}$$

where k_r is the rate constant for conformational change of the adsorbed protein.

Combining (24) and (27) by noting that at steady state $R_{ad} = R_{des}$, they get

$$A \, \Sigma \, k_a^{(i)} C_i(t_i) = k_d C_s - (1 + k_r t^{n-1}) \tag{28}$$

at $t \to 0$, this reduces to the Langmuir equation for reversible adsorption; as $t \to \infty$, the process becomes fully irreversible (see Fig. 19 which will be discussed later). The model reconciles essentially all of the physical evidence according to Soderquist and Walton [72], although several of its assumptions will be questioned later.

Sevastianov et al. [73,74] have developed a model which considers the effect of surface heterogeneity on the adsorption process. They define "centers of irreversible adsorption", labeled P, and "centers of irreversible desorption", labeled D. They argue, in agreement with Soderquist and Walton, that desorbed material is conformationally altered and thus cannot readsorb — hence desorption is irreversible. The results of this model are given as Fig. 14, taken from Ref. [73]. The model also includes the case where adsorption may be transport limited. The model fits commonly observed adsorption data, including the "overshoot" phenomenon (Fig. 14, top) (discussed in Ref. [72]) to be discussed later.

Although these models are ambitious, complex, and tend to fit experimental data

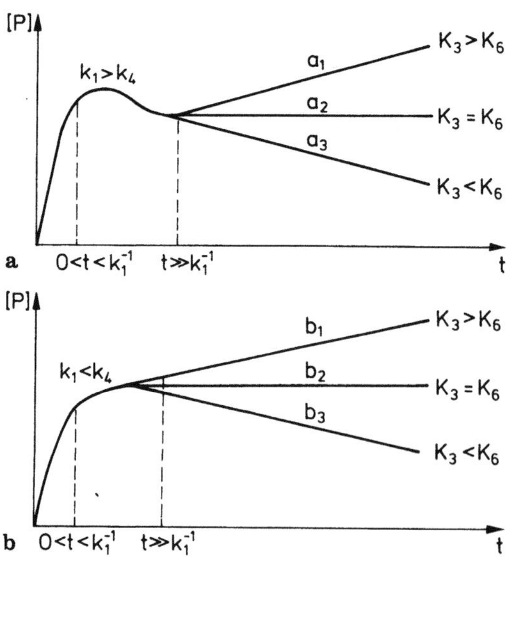

Fig. 14 a and b. Results of the Sevastianov, et al. kinetic model for protein adsorption [73,74]; k_1 is an adsorption rate constant [s^{-1}]; k_2, k_4, k_5 are "velocity" constants in [M/s]; and K_3 and K_6 are transport constants in [kg/M^2-s]. Curves a_1, a_2, a_3 (top) are for $k_1 > k_4$, $C_1 > 0$, $C_2 < 0$; curves b_1, b_2, b_3 (bottom) are for $k_1 < k_4$ and for C_1, $C_2 > 0$ (from Ref. [73])

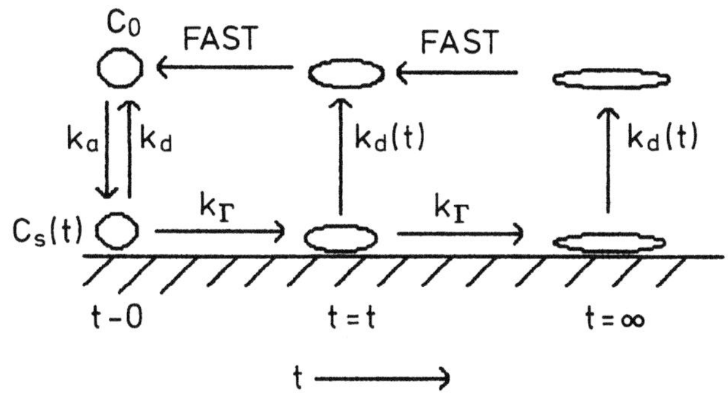

Fig. 15. Suggested general kinetic model for protein adsorption in the absence of any covalent bond formation or disruption. Any protein desorbed in a denatured state is assumed to rapidly renature in solution. If the surface is heterogeneous, then two or more such scenarios can be formulated, with appropriate account of the area fractions of each type of surface present

reasonably well, in many cases, some of the assumptions are not reasonable. Although proteins do indeed undergo time-dependent conformational changes on a surface, if those proteins desorb, they probably "renature" in solution, becoming indistinguishable from proteins in bulk solution. Protein denaturation is generally reversible *unless* covalent bonds have been formed or disrupted, which could happen if the surface is catalytically active or if proteolytic enzyme activity is present.

Therefore, a more general kinetic model is suggested in Fig. 15, based on the pioneering studies reviewed in Fig. 13. This is essentially a simplified form of the Soderquist and Walton [72] treatment, except we have assumed that any protein desorbed in a denatured state rapidly renatures in solution. This model could be combined with two or more classes of surface sites (heterogeneous surfaces) to treat more complex materials. To our knowledge, no modeling or data fitting to such a model is presently available.

4.4 Conformation Effects

We have already established that the protein adsorption process may result in significant conformational changes. In addition to adsorbed amounts and rates, the orientation and conformation of the adsorbed protein are critical (Fig. 16). Conformation refers to the secondary (α-helix, β-sheet), tertiary, and quaternary structures.

Certain orientations may make a specific site on the protein inaccessible to ligand, substrate, or antigen. For example, consider the adsorptive immobilization of a specific IgG for a solid-phase immunoassay. The procedure will be optimal if the Fab domains are free to bind antigen (Figs. 1 and 16).

If the protein's structure is changed due to the adsorption process and/or the new local micro-environment, then it is said to be fully or partially "denatured", meaning that its properties are no longer those of the native protein. Conformational changes can occur due to ligand binding (for example, hemoglobin), substrate binding (many enzymes), heparin binding (antithrombin III), and surface binding (Hageman Factor, fibronectin, albumin, etc.). Although there have been many hypotheses relating to protein adsorption conformational effects, there is little direct data available.

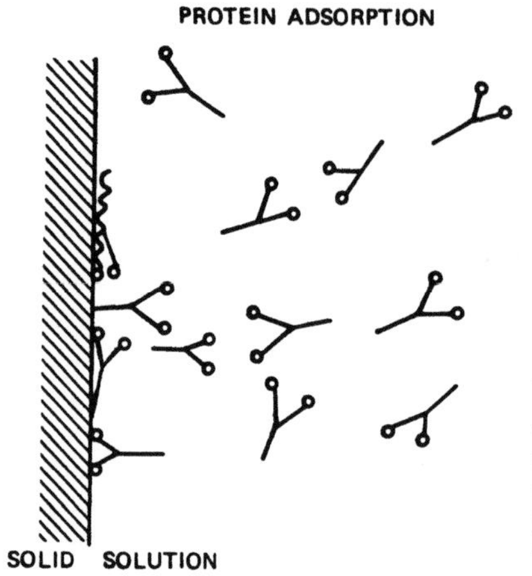

Fig. 16. The adsorption of an IgG molecule (shown for convenience as Y-shaped) may result in several orientations. The molecule may adsorb through one or both of the Fab ends or through the Fc tail — or lie on its "side". In addition to such *orientation* effects, the molecule may be conformationally altered or *denatured*

The surface itself may also be conformationally altered or "denatured", due to its interaction with the protein, particularly polymer surfaces which tend to relax or change in response to new environments [30].

We have already established that proteins are highly dynamic structures, which are constantly "sampling" different conformations due to local thermodynamic fluctuations. Some such transient conformations can be stabilized in new micro-environments, such as at a solid-liquid interface.

Many studies of proteins at air-solution interfaces have indirectly established that the adsorbed proteins undergo detectable conformational changes. Similar studies at solid-liquid interfaces are few. We review here only several key studies.

Morrissey [53] used transmission infrared spectroscopy to study protein adsorption onto silica particles in a heavy water (D_2O) buffer. By observing the shift in the amide I absorption band, he could deduce the fraction of protein carbonyl groups involved in bonding to the silica surface. He found that bovine IgG had a bound fraction of 0.20 at low bulk solution concentrations, but only about 0.02 at high solution concentrations. However, neither prothrombin nor bovine serum albumin exhibited a change in bound fraction with concentration. Parallel experiments with flat silica plates using ellipsometry showed that the IgG-adsorbed layers had an optical thickness of 140 Å and a surface concentration of 1.7 mg/m^2 at low bulk solution concentration — in concentrated solutions the surface amount was 3.4 mg/m^2 with a thickness of 320 Å (Fig. 17).

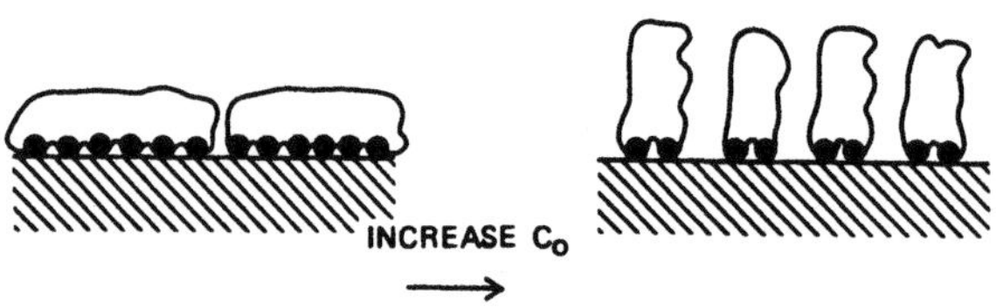

Fig. 17. At low solution concentration, the protein has no neighbors on the surface and thus can optimally adapt to the surface, maximizing the number of binding interactions. At high solution concentration, any one adsorbed protein is immediately surrounded by neighbors, minimizing the probability that it can conformationally adapt to the interface. This behavior leads to the differences in adsorbed amount and adsorbed protein thickness (determined by ellipsometry), as discussed in the text

Morrissey suggested that at low solution concentrations the adsorbed protein has sufficient time and "elbow room" to accommodate to its new microenvironment by conformational change resulting in significant hydrogen bonding to the silica surface. At high solution concentrations, the collision frequency with the surface is so high that an adsorbed protein has neither the time nor the room to optimize its interaction with the surface. This general interpretation is common in the air/solution field as well. Air/water studies show that the protein only needs to get a sufficient "foothold" on the surface to minimize the probability for desorption [3]. Once attached to the surface, the rest of the protein is "dragged" to the interface, thereby optimizing the interaction. The foothold need only be roughly 100–200 A^2, in contrast to the cross sectional areas of typical proteins, 1000–10000 A^2. Morrissey's high bulk concentration case may represent a minimum foothold.

Walton and his students have also pioneered the study of protein conformational changes, using fluorescence and circular dichroic spectroscopies [72,75,76]. Transmission CD spectra of Hageman Factor, adsorbed on a stack of quartz plates, suggested conformational changes upon adsorption. CD spectra of proteins eluted after different contact times with the surface show that the degree of conformational change is directly related to the contact time for periods of up to 10 days. It is now generally accepted that protein conformational change can be a rather slow process.

Brash has also used CD to study eluted proteins and finds large changes in α-helix content of fibrinogen, perhaps due to enzymatic fragmentation produced by the surface-induced activation of plasminogen to plasmin [77,78].

Transmission fluorescence studies of adsorbed Hageman Factor show changes, which can be interpreted in terms of conformational and other micro-environmental effects [76], although such studies must consider substrate effects on fluorescence [75].

Adsorption may lead to an increase or decrease in titrable groups. Titration data can thus be interpreted in terms of conformational changes.

Soderquist and Walton [72] showed an "overshoot" in adsorbed amount as a function of time and a kink or inflection in the isotherm at about half saturation (Fig. 18, see also Fig. 14). They proposed:

1. Within the first minute of contact, adsorption is rapid and reversible — a pseudo-

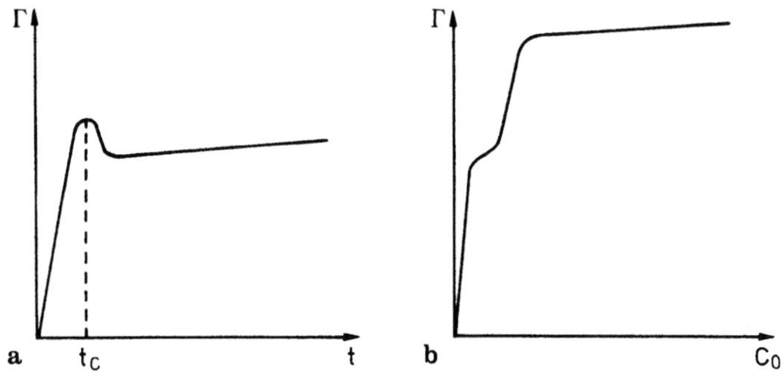

Fig. 18 a. A schematic of kinetic "overshoot" adsorption data, as discussed by Soderquist and Walton [72], Van Dulm and Norde [65], and others. See text for explanations. **b.** The "kink" at about half-saturation often observed in protein adsorption isotherms

equilibrium is present. The protein is adsorbed in a random arrangement at coverages less than 50%.

2. At coverages greater than 50%, surface transitions and ordering may develop which lead to more efficient packing and in increase in adsorbed amount, hence the kink in the isotherm.
3. Given sufficient time, adsorbed proteins undergo conformational changes which lead to increased surface interaction. During this process, proteins less optimally adsorbed undergo desorption, hence the overshoot in the time curve.
4. The desorption rate decreases with increasing residence time.
5. Desorbed protein may be permanently denatured, suggested by CD results.

Although this last point may not be general, the others are now accepted as rough "rules of thumb" for protein adsorption, though clearly each protein-surface-solution system is unique and may not exhibit all of the features noted.

Van Dulm and Norde [65] (in a study of human plasma albumin on negatively-charged polystyrene latices) showed a fast initial adsorption followed by desorption, probably due to conformational change of the adsorbed albumin, which induces the release of leass tightly bound protein. This result was observed at pH 4, where the albumin has a net positive charge and was not observed at pH 7.4, where it is highly negative.

The attenuated total reflection (ATR) Fourier transform infrared spectroscopic (FT-IR) studies of Gendreu, Jakobsen, and others [79] have the potential for direct determination of conformational changes during the adsorption process due to shifts in the infrared absorption bands. Sakurai et al. [80, 81], have used ATR-FTIR, as well as CD, to probe conformational changes upon adsorption.

Total internal reflection fluorescence (TIRF) spectroscopy has recently been applied by several groups [4, 39–42] and complete reviews are now available [4, 43]. The method can easily follow the kinetics of adsorption, using proteins labeled with extrinsic fluors, such as fluoroscein or rhodamine. The intrinsic UV fluorescence of tryptophan (Trp) can be used to follow adsorption. The UV approach has the advantage that the tryptophan fluorescence is sensitive to the local micro-environment and no label is required. The major disadvantage of the UV method is the UV photochemical changes which occur, although such changes can be minimized by working at low light levels.

The intrinsic UV fluorescence of proteins is dominated by the tryptophan indole rings. The absorption maximum is 280–290 nm with the fluorescence maximum ranging from 315–355 nm, depending on the local environment of the indole side-chains. Quantum yields range from 0.04 to 0.50; 0.10 is a common value. As the local environment polarity or dielectric constant increases, the fluorescence maximum shifts up to 355 nm, such as for an indole ring in water or buffer. Trp moieties in highly hydrophobic environments fluoresce at 315–320 nm. Thus the fluorescence emission maximum (and the quantum yield) provide indirect information as to the local environment of the Trp fluors.

Although a number of proteins of interest (human serum albumin, for example) contain a single Trp, most contain two or more. Thus the spectrum observed is the sum of all active Trp fluors, making it difficult to deduce the local environment of each fluor. Nevertheless, the UV fluorescence emission spectrum is useful in deducing orientation and/or conformation changes upon adsorption.

Fibronectin (Fn) adsorption from 0.05 mg/ml solution showed very different adsorption kinetics on hydrophobic and hydrophilic surfaces. The data for static and flow adsorption and desorption have been reported [82]. The quantity of interest here is the fluorescence maximum, which for Fn on hydrophilic silica is identical to that in bulk solution, suggesting no major conformational change upon adsorption. The Fn-silica surface interactions probably involve the charged groups on the surface of the molecule, most likely the highly positively-charged heparin binding regions of the molecule (pK = 8–9).

Fn adsorbed on hydrophobic silica, however, fluoresces at 326, suggesting a slight denaturation of the molecule. Fn interactions with the hydrophobic surface may involve some of the apolar residues in the protein interior, suggesting a partial denaturation. Clearly, studies on surfaces of a range of charge, polarity, and apolar character would be of interest.

Bovine albumin has been studied to date via intrinsic TIRF only on hydrophilic quartz [92,159]. The fluorescence maximum (1 mg/ml BSA in PBS) is 342 nm and shifts to 333 nm upon adsorption. These results suggest that the adsorption of BSA onto silica changes the conformation of the molecule such that the two Trp fluors are in a more hydrophobic environment.

Studies using radio-iodinated proteins in the TIRF apparatus permit a direct measurement of the amount adsorbed, thus allowing one to deduce if the quantum yield is also changed on adsorption [84].

Hydrophobic chromatography studies of proteins intramolecularly crosslinked to minimize unfolding suggest that "... the effective hydrophobicity of proteins is due in part to the extent to which buried hydrophobic residues are exposed by protein unfolding" (Ref. [83], p. 97).

Both Vroman [44] and Chuang [85–87] have shown that specific antibodies can be used as probes of adsorbed protein orientation or conformation. Chuang showed that fibrinogen adsorbed on Cuprophane surfaces could be measured quantitatively, using a ^{125}I-labeled antifibrinogen-IgG [85], however fibrinogen adsorbed on polyvinyl chloride "... was not readily accessible for reaction with ^{125}I-antifibrinogen-IgG."

Clearly specific antibodies, and particularly monoclonal antibodies, may be very useful in probing the properties of adsorbed proteins. Specific antibodies have been used to probe the structure of antigens in solution [88]. Consider the adsorption of a simple protein with a small number of reasonably well-defined epitopes (surface sites with antibody binding activity), as in Fig. 19. Clearly epitopes E and A are not accessible for binding, while B, C, and D would be sterically accessible. One could also envision a conformational change upon adsorption which produces an epitope

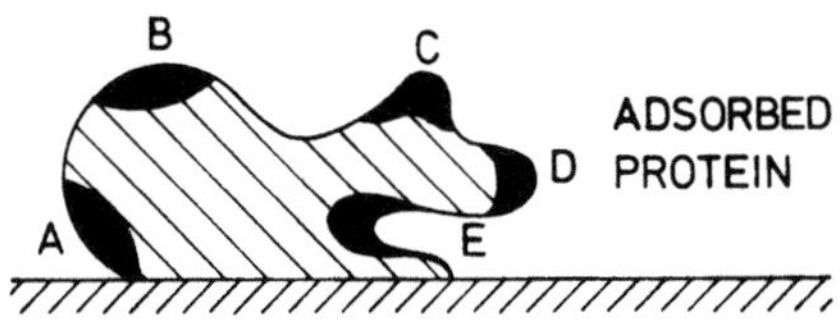

Fig. 19. Schematic of a protein with five different antigenic sites (epitopes). Each epitope may have one or more specific monoclonal antibodies. A set of such antibodies can be used to probe which epitopes are accessible or not, allowing the investigator to deduce the orientation of the adsorbed protein

normally unavailable. Antibodies to such "hidden" epitopes may be generated if the protein is first denatured in solution to expose the normally hidden site.

It is important to note that antibodies are very large molecules (Fig. 1) and may be sterically unable to bind to an adsorbed protein, even if the proper epitopes are indeed exposed to the solution (Fig. 19).

Chuang and coworkers observed that a specific antibody to thrombin showed very low reactivity with prothrombin in solution but very high reactivity with adsorbed prothrombin, suggesting that adsorbed prothrombin exposes a new binding site equivalent to one of the thrombin epitopes [87]. The reactivity of adsorbed prothrombin for the thrombin antibody was surface specific, suggesting that adsorbed prothrombin has different orientations and/or conformations on the surfaces examined. After a thorough study, Chuang concluded [87], "The data from this study appear to demonstrate that adsorption of prothrombin to artificial surfaces, such as PVC, had induced some conformational changes of the macromolecule that resembled antigenically the domain(s) of adsorbed thrombin . . . Monoclonal antibodies specific either to prothrombin or thrombin may be useful to pinpoint the exact domain(s) involved in such changes."

Chuang further cautions that "In solid phase immunoassay, it is generally assumed that antigens adsorbed to a surface, such as polystyrene microtiter dishes, will react with specific antibody in a manner similar to that antigen-antibody reaction in solutions such as occur in immune precipitation. However, our evidence and others [89] seem to point out that data obtained from solid phase immunoassays should be interpreted with caution since adsorption of a nonantigen to a polymer surface could render it immunoreactive to previously unreactive antibodies."

Clearly a set of monoclonal antibodies may help elucidate the nature of adsorbed protein orientation and conformation. Such studies are in progress by several groups.

4.5 Desorption, Exchange, Hysteresis

The adsorption of macromolecules is rarely an equilibrium process. Just as the properties of synthetic polymers are often dependent on non-equilibrium processes and relaxation phenomena [30], so do the properties of adsorbed proteins depend on time, metastable states, and hysteresis processes.

As the adsorption site density (Fig. 12) or total free energy of adsorption increases, one moves from the realm of "reversibility" to that of "irreversibility." As proteins can undergo conformational and orientational changes on a surface, they can optimize their interfacial interactions so as to provide the maximum free energy of adsorption. Such conformational alterations are relatively slow and hence very time-dependent.

Soderquist and Walton [72] showed that the desorption rate is a function of residence time. The adsorption process can be characterized in three stages:

1. Short times where adsorption is reversible; presumably little or no time is available for conformational changes.
2. At longer times where slow conformational changes occur, the process is semi-reversible and desorption occurs very slowly.
3. At long times where the conformational change is completed, adsorption is now irreversible, and desorption is improbable.

Those proteins which do not undergo any significant conformational change at a particular interface may not show any significant time-dependence.

Soderquist and Walton proposed that the kinetic "overshoot" (Fig. 18a) and the kink at around half-saturation (Fig. 18b) often observed in protein adsorption studies may be due to time-dependent, surface-induced transitions. The conformational changes probably result in some unfolding and an increased number of protein sites contacting the surface. Conformationally changed protein requires greater surface area — those proteins further along in the conformational change process "consume" surface at the expense of their neighbors, which are less tightly adhered to the surface. Therefore, the amount of protein adsorbed can go through a maximum as noted in Fig. 18. Van Dulm and Norde have made similar observations [65].

The isotherm ("equilibrium") case follows a similar argument (Fig. 18b). At low C_0, the surface is not saturated, and the adsorbed proteins are randomly oriented on the surface. At greater than 50% coverage, lateral interactions among neighbors become important, possibly producing an ordering of the adsorbed proteins, which can be thought of as a surface phase transition. The more ordered adsorbed layer requires less area per molecule, hence the surface has "room" for more proteins to adsorb — thus the kink in the isotherm (Fig. 18b). Direct evidence of two-dimensional protein crystallization and ordering is now available for the case of antibodies deposited on haptenated phospholipid monolayers [90].

Soderquist and Walton's [72] model of the protein adsorption process (discussed earlier, Fig. 13) reduces to the classical Langmuir (small-molecule) adsorption as $t \to 0$. As $t \to \infty$, their model predicts that adsorption is irreversible. A model which reduces to Langmuir adsorption at short contact time and to irreversible adsorption at very

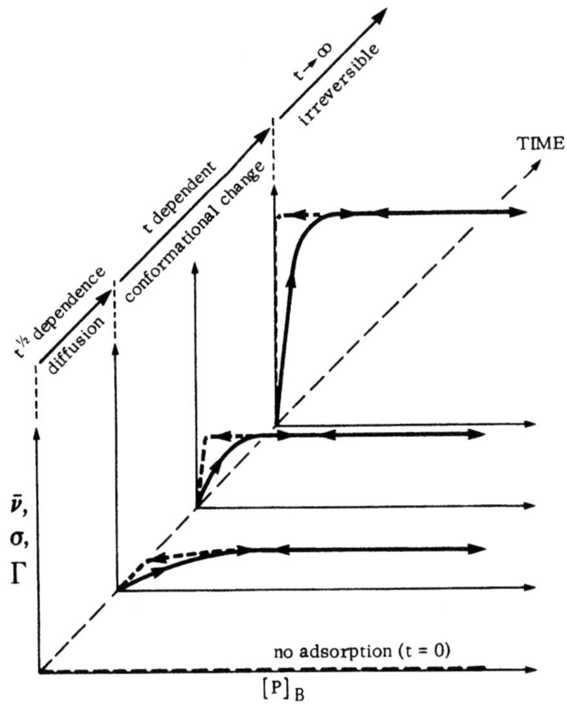

Fig. 20. Schematic adsorption "isotherms" with a constant surface site concentration ($[A]_s$ in Fig. 12 is here constant), but with adsorption time as a variable. At very short times, adsorption is diffusion controlled. At short times, the protein has insufficient time to conformationally adjust to the interface, thus adsorption can be reversible and of the Langmuir type. At longer times, conformational adjustments begin leading to the commonly observed semi-or ir-reversible behavior of protein adsorption. Other nomenclature same as Fig. 12

long contact times is certainly consistent with much of the data in the literature. Walton and Koltisko [91] have examined both protein stability and surface-induced, time-dependent conformational changes. Such treatments deserve to be more fully developed and investigated.

The tendency for an adsorbed protein to undergo conformational change is protein- and interface-specific, as well as time-dependent. Total internal reflection fluorescence (TIRF) studies on IgG adsorption and desorption on hydrophobic and hydrophilic surfaces as a function of residence time show clearly both the time-dependence, as well as the surface-dependence, of desorption [92].

Figure 20 summarizes the situation in schematic form. Assuming a constant surface binding site concentration (fixed surface properties), we examine the amount bound as a function of time and bulk solution concentration. At very short times and particularly at low bulk concentrations, adsorption is diffusion-controlled and shows a $t^{1/2}$ dependence. At longer times the sluggish conformational changes begin to become important. A significant hysteresis now begins to appear on the adsorption-desorption isotherm. At long contact times the conformational adjustments are complete, adsorption-free energy is maximized, and adsorption is irreversible, producing maximum hysteresis.

Clearly, proteins with a high concentration of binding sites on their surfaces may adsorb in the proper orientation, resulting in a multipoint attachment and high adsorption-free energy, even without any conformational adjustment. Thus, hysteresis can be present in some systems even at very short contact times. Proteins with very stable tertiary or quaternary structures may not show significant time-dependence due to the low probability for conformational change.

It is commonly observed that protein desorption can be very slow or even non-existant, but protein exchange can be rapid. This "anomaly" has been pointed out many times [93, 94]. Fortunately Jennissen's studies [95–97] and the arguments and work discussed in this chapter lead to a reasonable explanation.

Consider the cartoon in Fig. 21. Imagine some adsorbed proteins with multiple

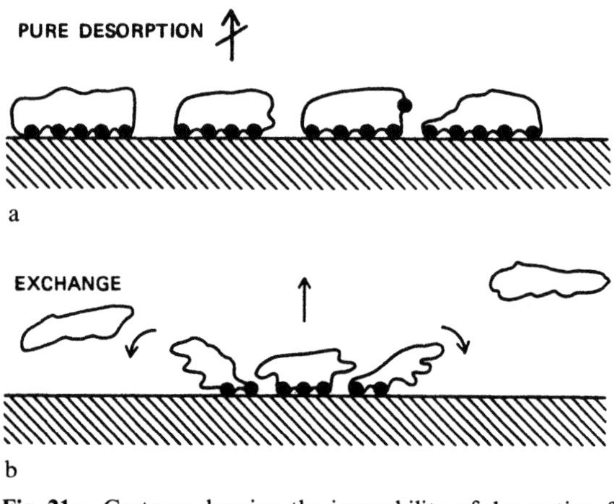

Fig. 21 a. Cartoon showing the improbility of desorption for a protein adsorbed by 4 or 5 binding sites; **b.** exchange of adsorbed protein with protein in the bulk solution is probable, however, due to the fact that as one foot releases, a foot from a different protein may attach. See text for further discussion. Explanation due to Jennissen [95–97]

binding sites. Call each protein binding site a "foot". Statistically, a protein foot may lift off every now and then. Clearly, all feet must be lifted off in order for the protein to desorb. This is statistically improbable, thus desorption cannot occur. We say the protein is irreversibly bound. Exchange is a different matter. Now we have proteins present in the local environment, diffusing to and colliding with the surface. If one of these proteins should statistically place a foot on the space vacated by a "desorbing" foot of an adsorbed protein, then a "competition" is present. If the new protein then puts several feet down and becomes anchored, it may indeed induce the removal of the first protein. Therefore, one protein can, in essence, be lifted off by a number of proteins adsorbing and attempting to accommodate with the surface. This is a very schematic and simplistic explanation, but it appears to at least qualitatively explain the situation.

The adsorption of polydisperse synthetic polymer shows similar behavior in that the average molecular weight of the adsorbed polymer increases with time — high molecular weight material is more tightly bound (more "feet" attached), thus inducing desorption of the low molecular weight molecules (see Fig. 22).

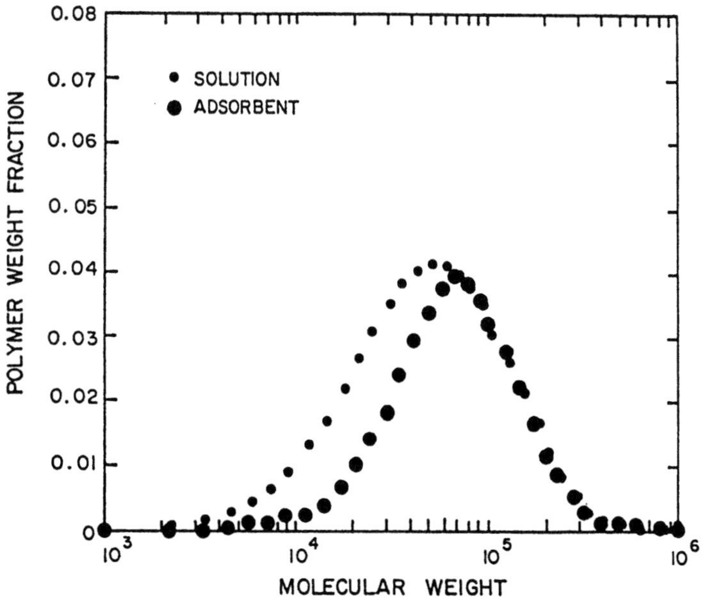

Fig. 22. Molecular weight effect of synthetic polymer adsorption. Molecular weight (MW) distribution of poly(vinyl chloride) in solution and in the adsorbed layer at equilibrium. Note that the adsorbed material has a higher average MW than the bulk solution (from Ref. [100], p. 120)

This is probably the basis of the complex protein exchange processes observed by Vroman and coworkers [98]. The more feet and energy per foot (total adsorption-free energy), the lower the probability that exchange can occur. Hence in a competitive adsorption process, one finds that those molecules present in the highest amounts and of smaller size bind first (due to diffusion and collision arguments) and are then exchanged by the more strongly adsorbing components. Vroman and Adams have shown [98] for plasma proteins adsorbing on glass and certain metal oxides that the exchange hierarchy is: albumin < gamma globulin < fibrinogen < fibronectin < Factor XII (Hageman Factor) < high molecular weight kininogen (HMWK) (albumin adsorbs first).

4.6 Competitive Adsorption — the "Vroman Effect"

The adsorption of protein from single component solutions is qualitatively understood, although a quantitative understanding and models or theories with predictive character are not yet available. *If* the structure and solution properties of the protein are known and *if* the solid-buffer interface properties are known, then by careful examination of the "surface" of the protein (ideally via molecular computer graphics), we can indeed predict what orientation of the protein is "preferred" on that particular surface.

Recently, considerable progress has been made on the calculation of electrostatic and hydrophobic interactions in biochemical systems [189-191]. We can expect such calculations to become common for protein-solid surface interactions. Thus we can expect approximate values for the adsorption energy in selected systems to appear in the near future. The problem of time-dependent conformational adaptation of the protein to the surface (and vice versa) will be much more difficult. Initially, we will have to resort to crude measures of the structural stability of a protein, such as the temperature at which thermal denaturation occurs, the urea molar concentration for solution denaturation, etc. One or more of the models given in Fig. 13 should apply.

But what happens if we have two or more proteins in solution? Clearly there will be a competition and the resultant surface concentration of the two proteins at some time, t, will be a complex function of protein, surface, and solvent properties.

The general problem involves, for example, three different proteins (A, B, C) in solution and interacting with a surface. The key variables are bulk solution concentration (C_A, C_B, C_C), diffusion coefficients (D_A, D_B, D_C), and the adsorption free energy, which is a function of time, $\Delta G(t)$ for each protein. The free energy of adsorption at initial contact, ΔG_0, is related to the final free energy state, ΔG_∞, through the "denaturability" of the protein. The denaturability is a function of the stability of the protein in solution, the interactions available at the solid surface, and the surface occupancy (fraction of sites occupied). The problem is to take these qualitative concepts and develop a quantitative model useful for predicting adsorption.

Although there have been many experimental studies of protein adsorption from binary and trinary mixtures, little or no modeling has been attempted.

Beissinger and Leonard [71] modeled the competitive adsorption of two proteins, albumin (A) and IgG (G), permitting each protein to occupy part of the surface in two different states, and also allowing for desorption of each protein from each of its two allowed states. The model uses 12 adjustable parameters, making it difficult to apply for predictive purposes.

Adsorption from plasma is considerably more complex, as literally hundreds of different proteins are present [99] — all competing for the surface.

Vroman has shown by antibody methods that plasma interactions with solid surfaces result in a hierarchial adsorption process [98]. The high concentration proteins dominate the surface at short times due to the higher collision rates. As time passes various exchange processes occur and proteins with higher surface affinities dominate the surface. Finally at very long times only the highest affinity proteins are present on the surface, even if their bulk solution concentration is very low.

This effect is well-known in synthetic polymer adsorption and results in low molecular weight species (fast diffusion) adsorbed initially, but the high molecular weight fractions are preferentially adsorbed at very long times (higher adsorption free energy) (see Fig. 22).

Adams, et al. [101] used a curved disc on a flat surface to study the effect of solution volume at constant surface area on competitive plasma protein adsorption. Although the experiment was qualitative, it elegantly demonstrated the importance of exchange, bulk solution concentration, and surface-volume ratio on competitive adsorption [98].

Recently Horbett [102] and Brash and ten Hove [103] have quantitatively demonstrated the "Vroman effect" in a series of experiments studying competitive adsorption of fibrinogen, albumin, IgG, and hemoglobin from diluted plasma.

The adsorption values at 5 minutes onto glass and polyethylene as a function of plasma dilution are given in Fig. 23. Fibrinogen adsorption on glass is maximal at about 1.5% plasma. From 0 to 1.5% plasma, a "typical" protein adsorption isotherm is observed. On polyethylene, the result for fibrinogen is similar. Albumin adsorbs in high amounts on the hydrophobic polyethylene surface showing a "normal" concentration dependence. The kinetics of fibrinogen adsorption show clearly that fibrinogen does indeed adsorb at the high concentration, but is removed or exchanged within a minute or so — exactly what Vroman observed many years ago.

Other workers have observed concentration-dependent competitive adsorption, including Grinnell and Feld [89] (fibronectin), and Breemhaar et al. [104] (fibrinogen, IgG, albumin).

What are these high affinity plasma components which compete so effectively for certain surfaces? Vroman suggests high molecular weight kininogen (HMWK), based on studies with HMWK-deficient plasma [98]. Breemhar et al. [104] suggest it could be a lipoprotein. Another explanation may lie in Brash's observation [77,78] that plasminogen can be activated by contact with glass to plasmin, which can then degrade adsorbed fibrinogen. Clearly there are many possibilities.

Clearly studies are needed which monitor the adsorbed protein population from plasma as a function of time. Only two methods come to mind:

1. Elute all adsorbed material from a sample and then do an ultrasensitive electrophoretic separation;
2. Use a large set of specific labeled antibodies to specifically and quantitatively bind with the adsorbed proteins.

Both approaches are fraught with problems.

5 Adsorption Based Biocompatibility Hypotheses and Correlations

5.1 Surface Charge

The surface charge concept goes back to the original observations that blood clots more rapidly in a glass tube than in hydrophobic glass or plastic tubes. This difference

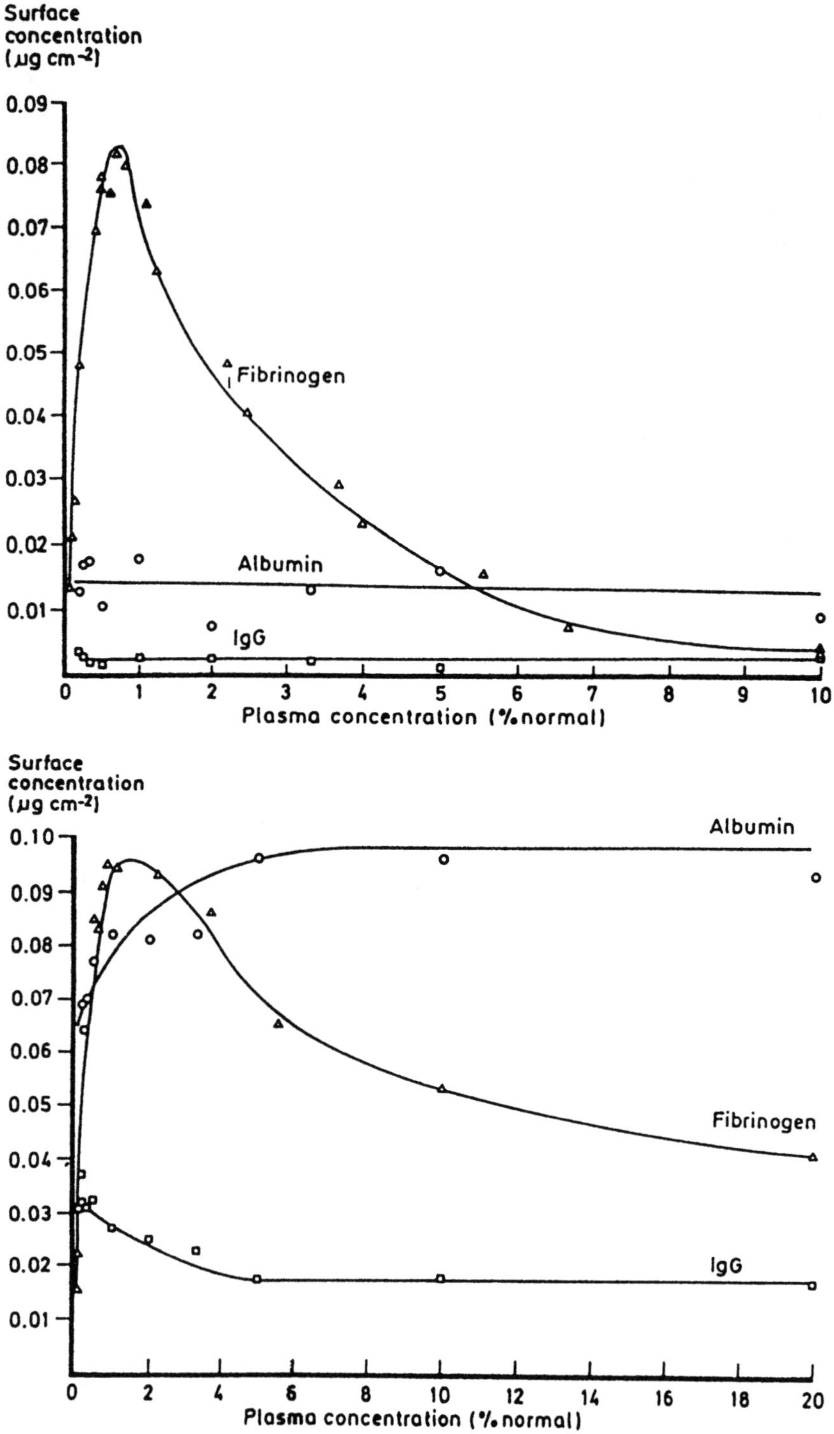

Fig. 23. Brash and ten Hove's results on the adsorption of three plasma proteins from diluted plasma as a function of total plasma concentration. Up: adsorption on glass showing a maximum adsorption of fibrinogen at about 1% plasma; Down: adsorption on polyethylene; plasma was diluted with isotonic Tris, pH 7.35. Adsorption time was 5 minutes (reprinted from Ref. [103])

was attributed to the negative surface charge of glass and is still quoted in hematology textbooks as the mechanism for surface-induced activation of coagulation. With the discovery of Hageman Factor and the intrinsic contact activation mechanism of coagulation, the negative surface charge theory was given some credibility. It was shown that Hageman Factor could indeed bind to negatively-charged surfaces: quartz, glass, and various silicate minerals. The counter theory argued that negatively-charged surfaces were more blood compatible because heparin, a common anti-coagulant, was highly negatively charged. Indeed, early studies by Lovelock and Porterfield [105] on the sulfonation of polystyrene to produce sulfonic acid groups analogous to those on heparin showed that such surface increased static blood coagulation times.

Sawyer's pioneering studies to measure the surface potential of the vascular surface by electrokinetic methods demonstrated that the vascular surface was negatively charged [106]. The mechanism suggested is that plasma proteins are generally negatively charged and therefore are repelled from a negatively-charged surface. This simplistic approach was very satisfying and dominated the blood compatibility field for several decades, despite the contradiction that Hageman Factor was known to be activated by negatively-charged surfaces.

It is now known that negatively-charged proteins are not repelled from negatively charged surfaces for a variety of reasons. One is that although a protein may exhibit a negative charge it may have localized regions or domains where negative charge is not present, where positive charge may be present, or where no charge is present. We demonstrated earlier in this chapter that proteins can adsorb to surfaces by a variety of interaction mechanisms. It is important to note that under normal physiologic conditions, the negative charge is screened by counterions beyond about 10 Å. By the time a protein is within 10 Å of the surface, specific intermolecular interactions are already dominant and gross electrostatic repulsion is not a significant effect.

Clearly, proteins can adhere to surfaces by electrostatic mechanisms, particularly at low ionic strength where the electrostatic field of the surface and the protein is much more extended. Indeed, this is the basis of ion exchange chromatography, so widely used for the separation, purification, and characterization of proteins. However, by the time one reaches the 0.15 M salt concentration of the physiologic environment, general electrostatic processes are no longer dominant.

5.2 Interfacial Energetics

With the wide availability of synthetic plastics in the late 50's and early 60's, there was considerable interest in relating the surface properties of plastics to their blood interactions. The only method of measuring a surface property was via wettability, i.e., contact angle measurements. Zisman and coworkers developed the critical surface tension concept [107], which permitted researchers to obtain empirical measures of the surface energy of polymeric materials. This approach was widely applied by Baier and eventually led to the hypothesis that surfaces with a critical surface tension in the range of 20–25 dyne/cm have optimal blood compatibility [108]. Such surfaces do indeed adsorb proteins. Baier argues that the critical surface tension of passive adsorbed protein films, and indeed of the vascular intima itself, is in the range of

20–25 dynes/cm. No satisfying mechanism has been proposed for this correlation. One suggestion is that the apolar component of the surface tension of water is 22 dynes/cm, within the range postulated by Baier.

An important hypothesis related to interface energetics is that of Nyilas, who said that the free energy of adsorption basically drives conformational change [132]. He probed this approach by measuring the enthalpy of adsorption using micro calorimetry and attempted to relate surfaces with low heats of protein adsorption with increased blood compatibility.

Shortly after the development of the critical surface tension concept Fowkes [109] and Girifalco and Good [110] developed means to estimate the surface-free energy and the surface-free energy components of solids by judicious application of probe liquids for contact angle studies and by intermolecular interaction approximations. These developments led Lyman and others to attempt to correlate the surface-free energy of polymers with their blood compatibility [111]. Lyman argued that as the surface-free energy increases, it increases the probability for protein binding and activation, such as by the Hageman Factor mechanism, and thereby decreases blood compatibility.

Lyman's ideas were extended by Andrade who argued that the governing parameter is not the surface-free energy of the polymer, but the interfacial-free energy of the polymer-water interface [112]. He argued that as the solid side of the interface begins to look more and more like water in energetic terms, there is a decreased driving force for protein adsorption and denaturation. This approach provided a semi-quantitative rationale for the rapidly developing interest in hydrogels as blood compatible surfaces.

With the development of neutral hydrophilic methacrylates in Prague, originally for contact lens applications in the early 60's [113], considerable interest was generated in the application of these materials in the cardiovasular environment. The qualitative argument was that such soft, water-rich surfaces must be relatively non-traumatic to proteins and cells. The development of neutral hydrophilic polysaccharide-based particles for protein chromatography in the late 60's provided evidence that such surfaces do indeed show minimal binding of proteins.

Basically, the interfacial-free energy hypothesis said that as the interfacial-free energy goes to zero, the driving force for protein adsorption goes to zero, and adsorption cannot occur [112]. If adsorption cannot occur, no mechanism is present by which to activate coagulation, and therefore the surface is blood compatible. Unfortunately, the contact angle methods and approximations for deducing interfacial free energy could not discriminate between materials ranging from about 40% water, such as poly(hydroxyethyl methacrylate), to over 95% water, such as the lightly cross-linked agaroses [114]. Thus the hypothesis could not be rigorously tested. Also, the highly hydrophilic surfaces required were not always stable or longlived in the cardiovascular environment.

5.3 Protein Passivation

In the mid to late 60's, considerable interest in the pre-adsorption of proteins evolved. The basic idea was that if one could saturate the surface with a layer of proteins, then the surface would not be available for the binding and activation of Hageman Factor or other contact activation proteins. Basically, albumin-treated surfaces were resistant to platelet adhesion, while other proteins — especially fibrinogen — promoted adhesion. This work has been well reviewed recently [115–117, 192].

The result was that albumin passivation came into vogue for sometime. In fact, artificial kidneys and blood oxygenators were often treated with albumin solutions prior to clinical use [118, 119]. There is considerable evidence that such pre-treatment did indeed result in decreased platelet adhesion and activation for short periods, perhaps up to several hours, but that the effect was relatively short-lived.

Matsuda et al. have argued [127] that practical blood compatible surfaces, such as the polyether urethanes, function by tightly adsorbing a layer of protein, which is highly denatured but well adhered to the surface. A second protein layer deposits on the first, also well adhered and partially denatured. The process continues as in Fig. 24 until an adsorbed protein film, perhaps a micron thick, eventually develops. Inner layers of the film are very tightly adhered and stable. The outer reaches of the film are basically in reversible equilibrium with circulating proteins in the blood. The problem occurs when such a film de-laminates or de-adheres from the surface thereby exposing bare surface, which can then, depending on the competitive adsorption and cellular processes present, activate coagulation and related processes. This model is reminiscent of that developed by Moacanin and Kaelble in which they argue that optimum blood compatibility is related to strong bio-adhesion events [128].

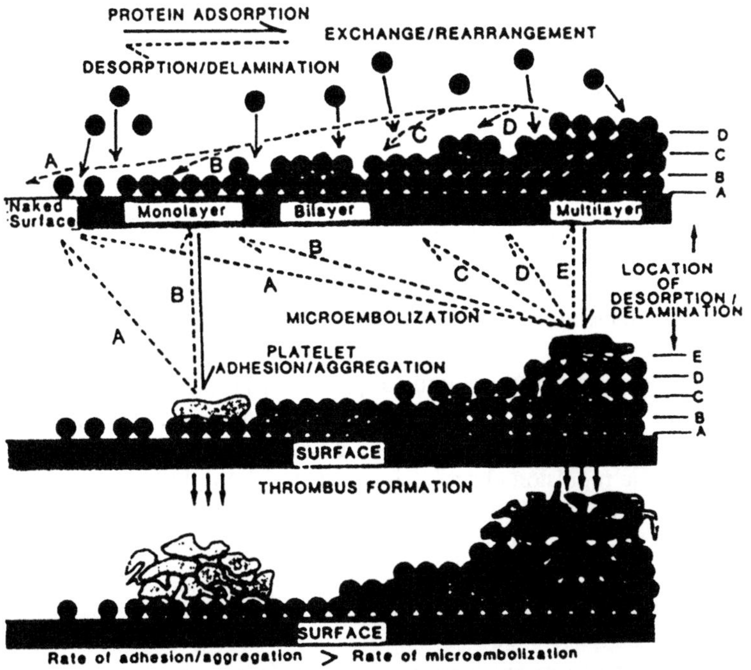

Fig. 24. Matsuda, et al.'s model of the protein adsorption/denaturation/aggregation/desorption/delamination process involved in the blood interactions of materials (from Ref. [127], p. 357)

5.4 Protein Resistent or Repulsive Surfaces

A current hypothesis, which is receiving considerable attention, is that one can indeed produce a surface which actively repels proteins and other macromolecules [123,124,133]. The basic idea is presented in Fig. 25, which shows that a neutral hydrophilic polymer, which exhibits considerable mobility or dynamics in the aqueous phase, can actively repel macromolecules from the interface by steric exclusion and interface entropy methods. This method has been well-known and applied in the field of colloid stability for many years [120]. The most effective polymer appears to be polyethylene oxide, probably because of its very high chain mobility and only modest hydrogen bonding tendencies [121–123].

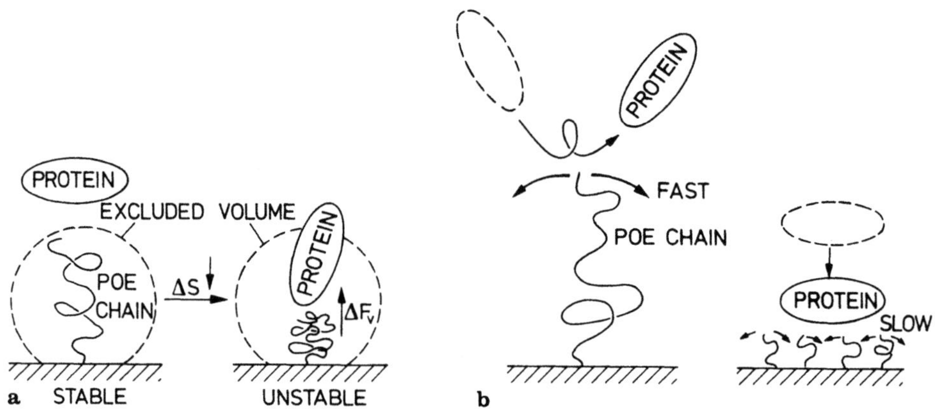

Fig. 25a and b. A protein resistant surface based on the steric repulsion argument commonly used in the colloid stability field [120]. The interaction between a polyethylene oxide grafted surface and a protein solution is shown. **a.** suggests an excluded volume or steric repulsion mechanism; **b.** the surface dynamics or polymer chain motion mechanism (from Ref. [133])

Merrill and Salzman have developed PEO soft segment polyurethanes and indeed have demonstrated minimal adsorption of blood proteins and minimal platelet adhesion on such surfaces [121]. Nagaoka et al. have studied various methacrylate copolymers with PEO side-chains of varying lengths and showed a direct correlation between minimization of platelet adhesion with increased PEO side-chain length and surface mobility [123,133].

Gregonis et al. have shown that PEO bound to quartz surfaces greatly minimizes the adsorption of proteins from plasma and from singly component protein solutions [122]. Bell and coworkers at Los Alamos have developed a theory of cell-cell adhesion based on the steric repulsion characteristics of the hydrophilic macromolecules present on cell surfaces [124].

5.5 Immunoglobulins and Complement

Within the last ten years or so there has been considerable interest in the activation of the Complement system by surface-induced processes. Immunoglobulin of the IgG or IgM class, adsorbed at a surface with the right orientation and spatial ordering and distribution, can activate Complement [125]. Certain hemodialyzer membranes, particularly of the cellulose type, are known to activate Complement via the alternate pathway, resulting in a transient, white cell depletion in the early stages of hemodialysis [125]. Sevastianov and Tseytlina recently reported the Complement activation properties of a variety of biomedical polymers [126]. There is therefore considerable interest on the adsorption of Complement components Cl and C3, as well as IgG, on surfaces, not only on cardiovascular materials, but also on intraocular lens implants, and at other sites of demonstrated Complement activity.

5.6 Other

Hoffman[129] and Baier [130] have reviewed most of the hypotheses and mechanisms suggested for blood compatibility in general and for the role of protein adsorption in particular. The safest statement one can make is that protein adsorption is indeed important in the blood compatibility process, in the compatibility of soft contact lenses, in the stability and acceptance of intraocular lenses, in the soft tissue foreign body reaction [131], and in virtually all situations where solid surfaces come into contact with physiologic environments.

It is also safe to say that, because of the great complexity of proteins, even of the simplest, most well-characterized proteins, such as insulin, lysozyme, and myoglobin, and because of the very wide range of proteins present in most physiologic environments, very simplistic hypotheses and mechanisms are generally not very applicable.

6 New Methods

6.1 Background

Techniques and methods for the study of protein adsorption have been well reviewed [4]. It is now generally recognized that it is not necessarily the type and amount of protein present at the surface which is most important, but rather the orientation and conformational state of those proteins. At present it is virtually impossible to predict the specific conformation of an adsorbed protein at a particular interface. The techniques used in the determination of protein conformation in solution or in the solid state do not usually apply to adsorbed proteins. Hence, the difference between adsorbed and bulk solution protein conformation has to be inferred indirectly.

Protein adsorption studies are performed either on high surface area material dispersed in a liquid phase containing dissolved protein, or on low surface area material, often flat, which is in contact with protein solution. Both approaches complement each other and can provide valuable information about adsorbed protein layers.

In the case of adsorbents with high surface area, changes in protein bulk solution concentration before and after adsorption are usually large enough for independent determination of the amount of protein adsorbed, either via solution depletion measurements or directly at the surface after the solid phase has been separated from the solution phase. Such studies are usually performed using various colorimetric methods of quantitative protein analysis or by measuring the radioactivity of proteins labelled with ^{125}I, ^{14}C, or ^{3}H. In some cases, labelling *per se* cause changes in protein adsorbability to a certain surface [134], producing different surface concentrations of labeled and nonlabeled protein and erroneous results. Similar problems apply also to protein labeled with fluorophores, such as fluorescein, rhodamine-B, and other dyes.

Due to the importance of the blood coagulation process, the adsorption of protein is often studied in buffer solutions at conditions similar to "physiological" (pH = 7.4, 0.15 M NaCl, 37 °C). Analysis of the adsorption process starts with the protein surface concentration presented as a function of equilibrium protein solution concentration, i.e., the adsorption isotherm. Most protein adsorption isotherms display a well-defined plateau in the dilute concentration range ($C_b < 1$ mg ml^{-1}). The adsorbed amount is usually up to the equivalent of a close packed monolayer, indicating an absence of multilayer adsorption. The adsorption isotherm is often claimed to be a "Langmuirian type", although the premises of Langmuir adsorption are rarely fullfilled, almost never investigated (with the exception of dilution effects), nor experimentally confirmed in the protein adsorption. The adsorbed amount usually changes only slightly with dilution, indicating apparent irreversibility [135]. However, it has been shown that proteins on the surface can exchange both with the proteins from the solution [136], as well as laterally with the neighboring protein molecules via surface diffusion [137]. The shape of the adsorption isotherm depends on experimental parameters; it can provide crude information about protein-adsorbent interaction. The initial isotherm slope reflects the affinity of protein towards the surface; the (apparent) adsorption isotherm plateau values can be related to the molecular dimensions of the adsorbed protein molecule. In some cases, the adsorption isotherms were found to display kinks which are thought to reveal distinct steps of interfacial protein rearrangement as the concentration of protein in the solution increases [138–140]. All of these effects were discussed earlier (Sect. 4).

In most cases, the adsorbing surface is tacitly assumed to be completely inert and nonresponsive to protein attachment. This assumption may be valid for surfaces of crystalline material but in the case of "hairy" polymer surfaces, changes in the polymer surface conformation due to the presence of adsorbed protein may be expected.

The major advantage of protein adsorption studies on high surface area materials is that changes of some extensive properties which accompany the process of adsorption are large enough to be directly measured: heat of adsorption through microcalorimetry [141], uptake or release of small ions by a combination of electrokinetic methods and titration [142], thickness of adsorbed layer or an increase of the volume fraction of solid phase by a hydrodynamic method like viscometry [143]. Chromatographic-like analysis can also be applied to protein adsorption [144].

For many areas of interest the most valuable information describing the interactions between a protein and a surface is conformational change of the adsorbing protein molecule as it passes from solution to the interface. Low surface area samples in combination with some form of spectroscopic method are generally used in the evaluation of protein conformation. Recent advances in this area warrant a more detailed description of the experimental approaches.

6.2 Spectroscopy by Evanescent Surface Waves [92, 154, 155]

As a rule these methods are based on the concept of evanescent surface waves caused by total internal reflection at a solid/liquid interface. In order to understand the so-called "evanescent" spectroscopy, a simplified theory of total internal reflection is given below. More complete treatments can be found in various specialized monographs and reviews [145-147].

An electromagnetic wave reflection at an interface forms a standing wave, due to superposition of the incident and reflected waves. This also occurs in the case of total internal reflection, where the standing wave forms in the optically more dense medium (medium 1). The nature of the standing wave is a function of both the optically more dense medium and the less optically dense medium (medium 2). In the familiar case of metallic reflection, the standing wave is aligned with its node at the reflecting metallic surface. In the case of the interface between two dielectric media, the electric field amplitude right at the interface, but in the denser medium, has a nonzero value. Since the boundary conditions for electromagnetic wave reflection do not allow any discontinuity in tangential field across the interface, the electric field amplitude at the interface in the less dense medium 2 will be equal to the electric field amplitude at the interface in denser medium 1. The solution of Maxwell's wave equation for total internal reflection shows the existence of a surface wave which propagates along the interface [148]. Its electric field amplitude decays exponentially into the less optically dense medium 2. In general,

$$E^{t} = E^{t,0} \exp\left(-\frac{z}{d_p}\right) \tag{29}$$

where E^t is the transmitted electric field amplitude at distance z normal to the interface; $E^{t,0}$ is the electric field amplitude right at the interface in less dense medium 2; and d_p is defined by:

$$\frac{1}{d_p} = \left(\frac{2\pi}{\lambda_2}\right)\left[\left(\frac{n_1}{n_2}\right)^2 \sin^2\theta_1 - 1\right]^{1/2} \tag{30}$$

where n_1 and n_2 are refractive indices of the optically more dense and less dense media, respectively; λ_2 is the wavelength of the electromagnetic wave in medium 2; and θ_1 is the angle of incidence of the electromagnetic wave measured from the interfacial normal. One can arbitrarily define d_p as the distance from the interface where the electric field amplitude decreases to e^{-1} of its interfacial value (often called the "depth of penetration of the evanescent wave"), i.e. $\exp(-z/d_p) = \exp(-1)$. An this case it can be shown that:

$$d_p = \lambda/2\pi(n_1^2 \sin^2\theta_1 - n_2^2)^{1/2} \tag{31}$$

The magnitude of $E^{t,0}$ can be calculated from Fresnel's law. If the electric field amplitude as a function of distance z is expressed per unit of incident electric field amplitude, $E^{i,0}$, of the perpendiculary polarized electromagnetic wave, then:

$$E_\perp^e = (E_\perp^{t,0}/E_\perp^{i,0}) \exp(-z/d_p) = (2\cos\theta_1)/[1-(n_2/n_1)^2]^{1/2} \exp(-z/d_p) \tag{31}$$

From an experimental point of view it is important to recognize that the profile of $(E_\perp^e)^2$ as a function of z is proportional to the profile of the intensity of electromagnetic radiation in the proximity of the interface in medium 2. Such a profile will determine the "surface sensitivity" of the evanescent wave; the "depth of penetration" is smaller if:
— there is greater mismatch between the refractive indices of two media,
— the wavelength of electromagnetic radiation is shorter, and
— the incident angle of electromagnetic wave is closer to its critical value.

Thus, choosing the experimental parameters one can design a sensing profile to a particular need. It has to be mentioned that in this simplified derivation of evanescent spectroscopy theory two implicit assumptions were made:
1. Both media are nonabsorbing, only real refractive indices were being used,[1] and
2. Any species adsorbing to the interface is not distinguished as an optically separate layer.

What is even more relevant to the present subject is that a thin dielectric layer of polymer can be situated between two media without grossly distorting the optical nature of the interface, i.e. total internal reflection would occur as if no polymer layer is present even if its refractive index is unmatched to both media. Such a layer will, however, affect the intensity of the evanescent wave and particularly its "depth of penetration."

One can distinguish between methods in which absorption of the evanescent surface wave in different wavelength regions is measured (these are often called "attenuated total reflection" methods), and methods which use the evanescent wave to excite other spectroscopic phenomena, like fluorescence and Raman scattering or light scattering. As the methods of conventional fluorescence spectroscopy have been shown to be exceptionally successful in studies of proteins and other biopolymers, their "evanescent" surface-sensitive counterparts will be reviewed first.

6.3 Total Internal Reflection Fluorescence (TIRF)

TIRF at solid/liquid interfaces was introduced by Hirschfeld [149]. Although this first use of TIRF was the study of bulk dissolved fluorescein in the vicinity of a fused silica-electrolyte interface, a number of advantages over the conventional transmission technique were demonstrated:

1. The intensity-concentration relationship was linear up to concentrations a hundred-fold higher;
2. Multiple total reflections increased the sensitivity of TIRF; and
3. Adsorption of dye to the surface increased the local concentration, enhancing the TIRF sensitivity.

[1] In this sense, the name "total internal reflection" is a misnomer since depending on the extent of the absorption of medium 2, there will be an electromagnetic energy "leakage" across the interface making rigorous critical angle definition not applicable. However, for weakly absorbing medium 2 the concept of total internal reflection is useful. The reader is referred to Ref. 146 and 147 for the derivations of evanescent wave equations for absorbing multilayered interfaces.

Further application involved collected of fluorescence from dansyl-labeled bovine serum albumin via TIRF optics [150], TIRF-immunoassay for specific dye-labeled antibodies binding from the solution to an antigen-coated surface [151], and a "virometer" — a optical sensor for viruses treated with a fluorescent probe bound to the virus nucleic acid [152,153].

In TIRF protein adsorption experiments, it is desirable to correlate the intensity of excited fluorescence with excess protein concentration at the interface. Such an adsorbed layer is often in equilibrium with bulk-nonadsorbed protein molecules which are also situated inside the "evanescent volume" and thus contributing to the overall fluorescence. Various calibration schemes were proposed, using external nonadsorbing standards [40,154], internal standard in a form of protein solution together with a type of evanescent energy distribution calculation [154], and independent calibration of protein surface excess [155]. Once the collected fluorescence intensity is correlated with the amount of adsorbed protein, TIRF can be applied in the study of various interactions between surface and protein.

Two different sources of fluorescence are possible:

1. Proteins containing amino acids like tryptophan and tyrosine which intrinsically fluoresce upon excitation in the ultraviolet range can be employed, hence, the name total internal reflection intrinsic fluorescence-TIRIF, or
2. Protein can be covalently labeled with an extrinsic fluor like fluorescein, rhodamine, etc.

Both approaches have distinct advantages and disadvantages: the first approach provides the possibility of recording intrinsic protein fluorescence emission and excitation spectra, which in turn can provide information about conformational changes in adsorbed proteins. It is also directly comparable with the methods of conventional fluorescence spectroscopy of proteins in solution. An interesting combination of TIRF and ^{125}I-labeled protein γ-detection was recently developed for determination of both the amount of protein and its fluorescence quantum yield in the adsorbed state [26]. It was shown recently that with appropriate apparatus the intrinsic fluorescence lifetimes of adsorbed protein can be determined [27]. Some proteins may show very weak fluorescence due to low quantum yield, have only a small number of tryptophanyl and/or tyrosinyl residues per molecule, and may be particulary photosensitive and unstable upon exposure to ultraviolet light [92].

The extrinsic probe approach is more suitable to competitive protein adsorption studies and to kinetic studies, provided that protein labeling by an extrinsic fluor does not influence protein adsorbability. The uptake of fluorescein-labeled albumin, γ-globulin and fibrinogen onto silicone rubber coated surfaces has been followed as a function of time and flow rate [39]; it was demonstrated that adsorption was diffusion limited [40]. A combination of total internal reflection with either fluorescence photobleaching recovery (TIR/FRP) or fluorescence correlation spectroscopy was described[137,155,157]. With this variation of TIRF, it is possible to determine surface desorption rates and surface diffusion coefficients without unnecessary perturbation of chemical equilibrium, like bulk protein concentration changes or changes in the composition of the buffer solution. Energy transfer between multi-labeled BSA (donor/acceptor pairs: dansyl/eosin and 4-chloro-7-nitro-2,1,3-benzoxadiazole/rhodamine) has found to decrease upon the adsorption of BSA, as detected by TIRF [158], a feature that can be interpreted as a conformational change of adsorbed BSA. Fluorescent probes, used in probing membrane protein properties, have recently been employed in the evaluation of conformational changes upon adsorption [159].

Fluorescence from labeled adsorbed protein has also been excited with the evanescent surface wave created by integrated optics. Both optical fiber [160] and flat rectangular waveguides [193] have been used. Interesting use of optical fiber as a remote protein sensor was demonstated; the excitation light was sent down the fiber whose tip was immersed in protein solution, evanescently excited fluorescence was collected by the same fiber and delivered to a scanning monochromator [160].

Changing the incident excitation beam angle and/or the angle of fluorescence observation provides other interesting information: a fluorophore concentration profile from the interface to distances which are within an order of magnitude of the wavelength used [161]. In the case of adsorbed protein, variable angle total internal reflection fluorescence (VATIRF) would not only give as a result an average thickness of the protein layer but its spatial distribution.

In conclusion, TIRF promises to be exceedingly useful in the study of protein-substrate interactions. It gives *in situ*, possibly remote, real-time information about protein adsorption-desorption parameters, conformational changes upon adsorption and hopefully, nanosecond time-resolved fluorescence lifetime information about adsorbed proteins [156].

6.4 Absorbance Spectroscopy of Adsorbed Proteins

Both conventional transmission absorbance spectroscopy and its evanescent counterpart (often called attenuated total reflection (ATR) spectroscopy) are based on measuring the light characteristics after it has propagated through the sample. Such techniques are classified according to the wavelength of the light used. Ultraviolet wavelengths are absorbed by all proteins. This feature has been used in the transmission absorbance mode in the evaluation of molar absorptivity of β-lactoglobulin films on the surfaces of quartz plates [162]. The absorption of visible light by rhodamine-BSA adsorbed to a number of stacked quartz plates was measured as a mean to calibrate extrinsic fluor TIRF experiments [155]. The use of attenuated total reflection both in ultraviolet and visible wavelength range for protein adsorption studies has to our knowledge not been reported, although a UV-TIR absorbance study of synthetic polymer adsorption has been reported [163].

ATR spectroscopy in the infrared has been used extensively in protein adsorption studies. Transmission IR spectra of a protein contain a wealth of conformational information. ATR-IR spectroscopy has been used to study protein adsorption from whole, flowing blood *ex vivo* [164]. Fourier transform (FT) infrared spectra (ATR-FTIR) can be collected each 5–10 seconds [165], thus making kinetic study of protein adsorption by IR possible [166]. Interaction of protein with soft contact lens materials has been studied by ATR-FTIR [167]. The ATR-IR method suffers from problems similar to TIRF; there is no direct quantitation of the amount of protein adsorbed, although a scheme similar to the one used for intrinsic TIRF has been proposed [168]; the "depth of penetration" is usually much larger than in any other evanescent method, i.e. up to 1000 nm; water absorbs strongly in the infrared and can overwhelm the protein signal, even with spectral subtraction applied.

6.5 Raman Spectroscopy of Adsorbed Proteins

Raman spectroscopy is another powerful method used for the study of solution protein conformation. Its application at the solid/liquid interface is conceivable [169], but to our knowledge it has not been applied to protein adsorption on polymer surfaces. One of the problems is that most proteins have a small Raman scattering cross-section. In order to enhance the weak Raman scattering signal, use of an evanescent streak propagating through a thin polymer waveguide layer with adsorbed protein has been demonstrated [193]. Total internal reflection optics were used to induce Raman scattering of 5 μm thick dry BSA films on a sapphire internal reflection element [170]. Another possibility to enhance the signal is to induce resonance Raman scattering of a protein absorption band; in the visible wavelength range this would apply mainly to proteins containing heme in their structures. Two other ways of enhancing the Raman scattering signal from surfaces were devised: use of surface plasmon excitation at the surfaces of thin silver films [171] and surface enhanced Raman scattering (SERS) by silver island films and finely dispersed silver colloids [172]. The Raman spectra of 75 Å thin phospholipid monolayers has been obtained by the first method [173]. SERS has been used to induce resonance Raman spectra from cytochrome-c and myoglobin adsorbed at a silver electrode [174]. The full potential of Raman scattering techniques applied to the adsorption of proteins has yet to be exploited.

6.6 Ellipsometry

Ellipsometry is another powerful tool in solid/liquid interface analysis. While evanescent spectroscopy has the capability of sensing the adsorbed protein molecules by interaction with the evanescent surface wave, ellipsometry "sees" a protein layer on the reflecting surface as a distinct optical medium. Ellipsometry is based on the calculation of the optical properties of the reflecting surface, given the change in the state of the polarization of the reflected light. Once the optical properties of the bare surface are known, the ellipsometric analysis can be applied to any dielectric film deposited on the surface. The exact relations between wavelength, optical constants, thickness and incident angle were obtained in exact form in the 19th century, but they could not be solved in the closed form. Early work on ellipsometry used only approximate equations, neglecting the higher order terms of the Taylor expansions of the Drude equations. The introduction of computers made routine analysis of the equations readily available. The use of ellipsometry in protein-surface interaction studies advanced rapidly after Trurnit [175,176] introduced an automated recording ellipsometer and measured the activity of the proteolytic enzyme, chymotrypsin, at a solid/liquid interface. Bovine serum albumin was deposited on the solid phase and depending on the condition used (ionic strength and pH) chymotrypsin was either removed or adsorbed to the albumin layer.

Studies of the role of protein-surface interactions in blood coagulation were done by Vroman [56]. The plasma proteins were adsorbed onto various hydrophilic or hydrophobic surfaces. Vroman showed that fibrinogen was an important component of the plasma protein layer adsorbed to the solid/liquid interface.

A complete review of the early work applying ellipsometry to biomedical problems is available [177].

The surface of silicon has optimum optical properties and the adsorption of proteins was followed on both hydrophobic and hydrophilic silicon surfaces [178, 179]. A fibronectin adsorption study showed, for example, that the adsorption is partially reversible on the hydrophilic surface, but much less so on the hydrophobic surface [179]. Interaction of antibodies with preadsorbed fibronectin layers suggest that fibronectin adsorbs in different orientation or conformation on the two surfaces [179].

Ellipsometry can follow the interactions between two types of biological macromolecules, the first of those two bound physically to the surface, the other acting from the solution. The binding of conconavalin A to adsorbed mannan [180] and of cholera toxin to adsorbed ganglioside [183] are examples. The adsorption of complement factors to an antibody-coated surface was monitored by ellipsometry and a modification of the same method was used for quantification of migration inhibition of human polymorphonuclear leucocytes [182]. Interaction of proteins and cells with affinity ligands covalently coupled to silicon surfaces has been also studied [183].

Recent development in the application of ellipsometry to protein/substrate interaction studies is due to the introduction of the "dynamic ellipsometer", capable of recording both analyser and polariser positions automatically with time, which significantly improves the time resolution of the method [184]. The Lorentz-Lorenz equation is usually applied in order to calculate the mass of adsorbed proteins [185]. In particular, molecular weight, molar refractivity and partial specific volume are needed to calculate adsorbed mass from the experimentally determined thickness and refractive index of deposited film. While in most of the previous studies the assumption was made that the refractive index of the adsorbed layer was constant, the experiments done by Cuypers showed that such an assumption is not justified [186]. The refractive indices of protein layers varied as a rule with the time of adsorption [187]. So far, no general rules emerged about the course of refractive index changes. The refractive index of fibrinogen adsorbed on hydrophilic chromium oxide surface, for example, remained almost constant ($n = 1.4$) while the thickness increased to 12 nm. On the contrary, the refractive index of fibrinogen adsorbing on hydrophobic chromium was found to change significantly with time of adsorption; on the onset ($t < 150$ s) it was 1.8, only to drop to 1.48 while the thickness of the layer continued to increase to 7 nm. From this point on the fibrinogen layer became optically denser and thinner. Final values for the fibrinogen layer deposited from 10 μg/ml solution (0.01 M Tris-HCl buffer, pH = 7.0) onto hydrophobic chromium surface were: thickness, $d = 3.5$ nm, refractive index, $n = 1.72$. Such changes directly reflect some conformational changes and packing or ordering of the adsorbing protein. It is, however, difficult to explain such a high value of refractive index of the protein layer. It was reported that the refractive indices of prothrombin layers adsorbed on two phospholipids (di-C14:0 phosphatidylserine and di-C18:0 phosphatidylserine) were 1.46 and 1.90, respectively [185]. In order to account for this discrepancy, the authors speculated that refractive index values higher than the values for the pure protein reflected an interaction between the protein and the adsorbing surface and that in order to compensate for higher refractive indices of the protein film, a decrease of the refractive index of adsorbing surface has to be introduced. This assumption cannot be proved experimentally, but it was supported by finding that the adsorption of protein onto more swollen phospholipid layers, (i.e. onto a layer with a lower refractive index) resulted in higher refractive indices of protein film [187]. Clearly, more detailed modeling and computer simulation studies would be helpful.

7 Some "Rules of Thumb" (See also Sect. 4.2)

7.1 Area Required For Initial Adsorption

At an air/water interface, the two-dimensional interfacial pressure (Π) can be easily monitored using an instrumented Langmuir trough. The initial adsorption rate at a clean surface is simply the rate of diffusion

$$\frac{dn}{dt} = K_a C_0 \tag{32}$$

where K_a is the adsorption rate constant. Assume that the interface is partially occupied and the interfacial pressure is Π. "In order for a molecule to adsorb, it must compress molecules already adsorbed against the interfacial pressure Π to create an area of interface, ΔA, equal to that required for the molecule to move into (Ref. [3], p. 289)."

The work required is $\int_0^{\Delta A} \Pi dA$ or $\Pi \Delta A$ if Π is approximately constant. Then

$$\frac{dn}{dt} = K_a C_0 \exp\left(\frac{-\Pi \Delta A}{kT}\right) \tag{33}$$

where k is Boltzman's constant, T is absolute temperature, and no desorption has been assumed (valid for $\Pi < 15$ mN/m). Equation 33 can be expressed as:

$$\ln\left(\frac{dn}{dt}\right) = \ln(K_a C_0) - \frac{\Pi \Delta A}{kT} \tag{34}$$

A simple linear plot of the data allows ΔA to be obtained. The results of a set of experiments (Table 5) are surprising. ΔA is independent of the size or molecular weight of the protein. Although the cross-sections of the proteins studied range from ~1000 to 10,000 A^2, ΔA is nearly constant at 100 to 200 A^2. Conclusion: "*. . . only a small portion of the protein molecule needs to enter the interface in order for adsorption to then proceed spontaneously* (Ref. [3], p. 290)." It is as if only a small "foothold" or "handhold" is required to stabilize the molecule against desorption. Now firmly planted at the interface, the molecule can optimize its interfacial interactions by time-dependent orientation and perhaps conformational changes. The size of the "foot" is obviously relevant to the exchange discussion in Sect. 4.5.

Table 5. Values of ΔA for various proteins (from Ref. [3])

Protein	Concentration ($g\,l^{-1}$)	ΔA $(Å^2)^{-3}$	Molecular Weight
Myosin	0.03	145	600000
Human γ-globulin	0.01	130	180000
Human albumin	0.02	100	70000
Ovalbumin	0.03	175	44000
Lysozyme	0.01	100	15000

An alternative interpretation of the data is that ΔA is proportional to the number of water molecules per protein participating in the adsorption and denaturation process and is related to water activity at the interface.

7.2 Electrical Potential Barrier

A similar experiment to that noted above can be performed, but now let the interface be populated by a molecular layer at constant Π and known interface electrical potential. A molecule adsorbing at such an interface must do work against the electrical potential barrier, as well as against the interfacial pressure. We get

$$\frac{dn}{dt} = K_a C_0 \exp - \left(\frac{(\Pi \Delta A + q\Psi)}{kT}\right) \quad (35)$$

where q is the charge on the adsorbing molecule and Ψ is the interface electrical potential (3). Table 6 and Fig. 26 present data on lysozyme adsorption into different monolayers. Clearly the *electrostatic effects are very significant.*

Table 6. Relative initial rates of adsorption of lysozyme into different monolayers[a]

Monolayer	Rate	ζ-Potential (mV)
Cephalin	34.5	−68.1
Polyglutamic acid	13.2	−51.5
Pepsin	5.4	−16.7
Serum albumin	4.5	−14.1
Octadecanol	5.0	Not measured
Trypsin	3.0	+3.2
Lysozyme	(1.0)	+12.2
Polylysine	0.3	+38.5

pH 6.5, I.S. = 0.01, charge on lysozyme ≃ +9 units.
[a] from Ref. 3

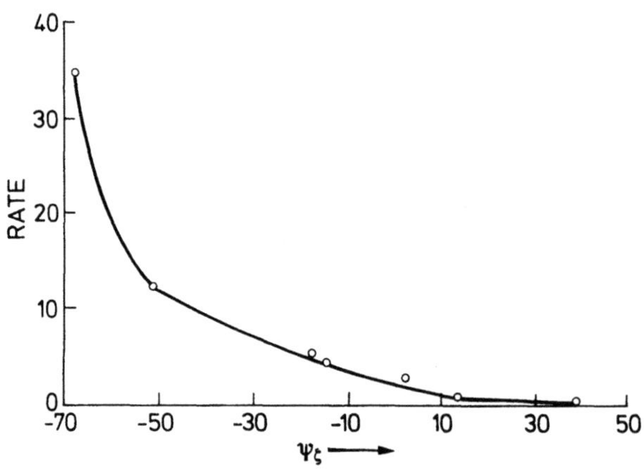

Fig. 26. Plot of the data of Table 6 showing rate of lysozyme adsorption at air/water monolayers of different zeta potentials. Refer to Eq. (35) and text for details (from Ref. [3])

Because of the large size of the protein and the ΔA required for initial adsorption, only a small region of the molecule need participate in the initial stage of the process.

It is well-known that *protein adsorption tends to be at a maximum at the isoelectric point* because the protein has zero net charge. Under such conditions, electrostatic barriers to adsorption are minimized.

Protein adsorption is dependent on the nature and concentration of the electrolyte. Norde and Lyklema have shown [61] that adsorption generally increases with decreasing charge on the protein and on the surface, although counterion effects can often override this generalization.

7.3 Hydrophobicity

The now classic studies by Norde and Lyklema and their detailed thermodynamic analysis (Sect. 4.2) have established that the *interaction between a protein and a surface increases with increasing hydrophobicity of the surface and increases with increasing hydrophobicity of the protein.* "Desorption from hydrophobic surface usually does not occur whereas proteins can often be removed from hydrophilic surfaces by exposure to extreme pH, high ionic strength, or by extensive rinsing [61]."

7.4 Diffusion and Mass Transport

Proteins have a low diffusivity: $D \sim 2$ to 6×10^{-7} cm^2/s. Therefore, the *initial stages of adsorption are generally diffusion limited* and the amount adsorbed is proportional to $t^{1/2}$. If the adsorption rate, dn/dt is:

1. $\sim 1/t^{1/2}$, diffusion is rate limiting;
2. $> 1/t^{1/2}$, then convection or other mass transport effect is present;
3. $< 1/t^{1/2}$, the adsorption is reaction limited; i.e., there are energy barriers to the adsorption process.

7.5 Time

As *contact or residence time increases, the protein tends to orientationally and conformationally adjust to the interface*, leading to stronger bonding, and greater irreversibility of adsorption.

7.6 Surface

Surface adsorption site energy and density are very important. Most *biomaterial surfaces have very high site densities*, making it difficult to study the mechanisms governing adsorption. Low site density surfaces are available. Heterogeneous surfaces, such as block copolymers and polymer blends, may have very unique adsorption properties. If one of the phases or domains tends to dominate the surface, it may act as a homogenous surface. If both phases are present on the surface, then two or more very different classes of adsorption sites will be present. Protein adsorption on each of the two phases can be very different [80, 81].

7.7 Protein

The protein's intrinsic properties (size, molecular weight, 3-D structure, surface site density, conformational stability) are all very important and must be fully characterized and understood in order to interpret adsorption data.

8 Summary/Conclusions

The general principles of protein adsorption are beginning to be identified and understood. New techniques and methods are now available which, together with well-established methods, allow one to thoroughly probe the adsorption process.

In the past, the solid surface and the solid/solution interface were often poorly characterized and poorly understood. A range of new concepts and tools are now available for the study of solid/solution interfaces [30]. A new monograph on protein adsorption is now available [4]. An international conference on protein and polyelectrolyte adsorption was recently held and proceedings will soon be available [188].

Protein surface structure and conformational dynamics are now much better understood. In the near future, we can expect extensive application of computer molecular graphics to better visualize and understand protein-surface interactions.

Very qualitatively — protein adsorption is roughly understood. The challenge now is to take what we know about proteins, surfaces, and adsorption and to begin to quantitatively model the protein adsorption process.

The future looks bright!

Acknowledgements: This work was supported by NIH grants HL-16921, HL-18519, and HL-29209 by the Faculty Research and Biomedical Support Grants, and by the Center for Biopolymers at Interfaces, Univ. of Utah. V. Hlady thanks the Fullbright Program for support and the R. Bošković Institute, Zagreb, Yugoslavia for a leave of absence.

9 References

1. Brash, J. L. and Lyman, D. J.: Chemistry of Biosurfaces *1*, Hair, M. L. (ed.), New York: Dekker, 177–232 (1971)
2. Vroman, L.: Blood, Natural History Press, 1966
3. Macritchie, F.: Adv. Protein Chem. *32*, 283–326 (1978)
4. Andrade, J. D. (ed.): Protein adsorption, New York: Plenum Press, 1985
5a. Vroman, L.: Nature *196*, 476-7 (1962)
5b. Vroman, L.: Thromb. Diath. Haemorrh. *10*, 455 (1963)
6. Dickerson, R. E. and Geis, I.: Structure and action of proteins, New York: Harper & Row 1969
7. Stryer, L.: Biochemistry, 2nd ed., San Francisco: W. H. Freeman Co. 1981
8. Schulz, G. E. and Schirmer, R. H.: Principles of protein structure, New York: Springer 1979
9. Dickerson, R. E. and Geis, I.: Hemoglobin, Menlo Park: Benjamin/Cummings 1983

10. Freifelder, D.: Physical biochemistry, 2nd ed., San Francisco: W. H. Freeman Co. 1982
11. Cantor, C. R. and Schimmel, P. R.: Biophysical chemistry, San Francisco: W. H. Freeman 1980
12. Bull, H. B.: An introduction to physical biochemistry, Philadelphia: F. A. Davis Co. 1964
13. Walton, A. G. and Blackwell, J.: Biopolymers, New York: Academic Press, 1973
14. Walton, A. G.: Polypeptides and protein structure, New York: Elsevier 1981
15. Tanford, C.: Physical chemistry of macromolecules, New York: Wiley 1961
16. Lyklema, J.: Affinity chromatography, Gribnau, T. C. J. (ed.), Amsterdam: Elsevier, 11–27 (1982)
17. Franks, F. and Eagland, D.: CR Crit. Rev. Biochem. *3*, 165–219 (1975)
18. Anfinsen, C. B.: Science *181*, 223–230 (1973)
19. LaBrecque, M.: Mosaic, 3–9 (May/June 1983)
20. Pace, C. N.: CRC Crit. Rev. Biochem. *3*, 1–43 (1975)
21. Chen, J. and Andrade, J. D.: Biomaterials *6*, 231 (1985)
22. Cooper, A.: Sci. Prog. *66*, 473–497 (1980)
23. Jardetzky, O.: Acc. Chem. Res. *14*, 291–298 (1981)
24. Feldmann, R. J.: Proc. Natl. Acad. Sci. U.S.A. *75*, 5409–5412 (1978)
25. Furie, B., Bing, D. H., Feldmann, R. J., Robison, D. J., Burnier, J. P., Furie, B. C.: J. Biol. Chem. *257*, 3875–3882 (1982)
26. Tucker, J. B.: High Technology 52–60 (1984)
27. Connolly, M. L.: Science *221*, 709–713 (1983)
28. Feldmann, R. J., Bing, D. H.: Teaching aids for macromolecular structure, New York: Taylor Merchant Corp.
29. Hansen, J., Andrade, J. D., and Ely, K.: J. Colloid Interface Science, submitted.
30. Andrade, J. D., ed.: Surface and interfacial properties of biomedical polymers, vol. 1, New York: Plenum Press 1985.
31. Davies, J. T. and Rideal, E. K.: Interfacial phenomena, New York: Academic Press 1963
32. MacRitchie, F.: Interfacial synthesis, V. I. fundamentals, Millich, F. and Carraher, C. E. Jr. (eds.), New York: Dekker, 103–138 (1982)
33. Gomer, R.: Scientific Amer., 98–109 (Aug. 1982)
34. Burghardt, T. P. and Axelrod, D.: Biophys. J. *33*, 455–468 (1981)
35. Michaeli, I., Absolom, D. R., Van Oss, C. J.: J. Coll. Interface Sci. *77*, 586-7 (1980)
36. Daugherty, R. L. and Franzini, J. B.: Fluid mechanics, New York: McGraw-Hill 1977
37. Ruzicka, J. and Hansen, E. H.: Flow injection analysis, New York: Wiley 1981
38. Levich, V. G.: Physico chemical hydrodynamics, New Jersey: Prentice-Hall 1962
39. Watkins, R. W. and Robertson, C. R.: J. Biomed. Materials Res. *11*, 915–938 (1977)
40. Lok, B. K., Cheng, Y-L., Robertson, C. R.: J. Colloid Interface Sci. *91*, 87 103 (1983)
41. Beissinger, R. L. and Leonard, E. F.: ASAIO J. *3*, 160–175 (1980)
42. Watkins, R. W.: TIRF for the study of protein adsorption, Ph. D. dissertation, Stanford University, Nov. 1976
43. Axelrod, D., Burghardt, T. P., Thompson, N. L.: Ann Rev. Biophys. Bioeng. *13*, 247–268 (1984).
44. Vroman, L.: Biocompatible polymers, Szycher, M. (ed.), Lancaster, PA: Technomic Publ., 81–88 (1982)
45. Adamson, A. W.: Physical chemistry of surfaces, 4th ed., New York: Wiley 1983
46. Kipling, J. J.: Adsorption from solutions of nonelectrolytes, New York: Academic Press 1965
47. Steinhardt, J. and Reynolds, J. A.: Multiple equilibria in proteins, New York: Academic Press, 1970
48. Van Holde, K. E.: Physical biochemistry, New Jersey: Prentice-Hall 1971
49. Marshall, A. G.: Biophysical chemistry, New York: Wiley 1978
50. Jennissen, H. P.: Biochem. *15*, 5683–5692 (1976)
51. Jennissen, H. P.: Z. Physiol. Chem. *357*, 1201–1203 (1976)
52. Hofstee, B. H. J. and Otillio, N. F.: J. Chromatog. *159*, 57–69 (1978)
53. Morrissey, B. W.: Ann. N. Y. Acad. Sci. *288* 50–64 (1977)
54. Weber, G.: Adv. Protein Chem. *29*, 1–83 (1975)
55. Weber, G.: Molecular biophysics, Pullman, B. and Weissbluth, M. (eds.), New York: Academic Press, 364–396 (1965)
56. Laiken, N. and Nemethy, G.: Biochem. *10*, 2101–2106 (1971)

57. Nemethy, G.: Subunits in biological systems, part C, Timisheff, S. N. and Fasman, G. D. (eds.), New York: Dekker, 1–90 (1975)
58. Laiken, N. and Nemethy, G.: J. Phys. Chem. *74*, 4421–4441 (1970)
59. Poland, D.: Cooperative equilibria in physical biochemistry, Oxford: Oxford Press, 1978
60. Levitski, A.: Quantitative aspects of allosteric mechanisms, Berlin: Springer-Verlag 1978
61. Norde, W.: Adhesion and adsorption of polymer, vol. 2, Lee, L.-H. (ed.), New York: Plenum Press, 801–826 (1980)
62. Lyklema, J.: Colloids and Surfaces *10*, 33–42 (1984)
63. Norde, W.: Colloids and Surfaces *10*, 21–31 (1984)
64. Walton, A. G. and Koltisko, B.: Biomaterials, Adv. Chem. Series #199, Cooper, S. L. and Peppas, N. A. (eds.), Washington DC: American Chemical Society, 245–264 (1982)
65. Van Dulm, P. and Norde, W.: J. Colloid Interface Sci. *91*, 248–255 (1983)
66. Norde, W. and Lyklema, J.: J. Colloid Interface Sci. *71*, 350 (1979)
67. Norde, W. and Lyklema, J.: J. Colloid Interface Sci. *66*, 257 (1978)
68. Koutsoukos, P. G., Norde, W., and Lyklema, J.: J. Colloid Interface Sci. *95*, 385 (1983)
69. Van Dulm, P., Norde, W., and Lyklema, J.: J. Colloid Interface Sci. *82*, 77 (1981)
70. Lundström, I.: Phys. Scripta *T4*, 5–13 (1983); see also J. Theor. Biol. *110*, 195–204 (1984)
71. Beissinger, R. L. and Leonard, E. F.: J. Colloid Interface Sci. *85*, 521–533 (1982)
72. Soderquist, M. E. and Walton, A. G.: J. Colloid Interface Sci. *75*, 386–392 (1980)
73. Sevastianov, V. I., Belomestnaia, Z. M., Zimin, N. K.: Artif. Organs *7*, 126–133 (1983)
74. Sevastianov, V. I., Tseytlina, E. A., Volkov, A. V., and Shumakov, V. I.: Trans. Am. Soc. Artif. Intern. Organs *30*, 137–141 (1984)
75. Walton, A. G. and Maenpa, F. C.: J. Colloid Interface Sci. *72*, 265–278 (1979)
76. McMillin, C. R. and Walton, A. G.: J. Colloid Interface Sci. *48*, 345–349 (1974)
77. Chan, B. M. C. and Brash, J. L.: J. Colloid Interface Sci. *84*, 263-5 (1981)
78. Brash, J. L. and Chan, B. M. C.: Abstracts, 8th Ann. Meet. Biomaterials Soc., 85, (April 1982)
79. Gendreau, R. M., Leininger, R. I., Winters, S., and Jakobsen, R. J.: Biomaterials, Adv. Chem. Series #199, Cooper, S. L. and Peppas, N. A. (eds.), Washington DC: American Chemical Society, 371–394 (1982)
80. Okano, T., Nishiyama, S., Shinohara, I., Akaike, T., and Sakurai, Y.: Polymer J. *10*, 223–228 (1978)
81. Sakurai, Y., Akaike, T., Kataoka, K., and Okano, T.: Biomedical polymers, Goldberg, E. P. and Nakajima, A. (eds.), New York: Academic Press, 335–379 (1980)
82. Iwamoto, G. K., Winterton, L., Stoker, R. S., Van Wagenen, R. A., Andrade, J. D. and Mosher, D.: J. Colloid Interface Sci., *106*, 459 (1985)
83. Mann, D. F. and Moreno, R. O.: Prep. Biochem. *14*, 91–98 (1984)
84. Reinecke, D., Van Wagenen, R. A., Andrade, J. D., and Smith, L. M.: Biomaterials 84, Trans. Soc. Biomaterials *7*, 276 (1984); also, Reinecke, D.: M. S. thesis, University of Utah, June 1985
85. Chuang, H. Y. K.: J. Biomed. Materials Res. *18*, 547–559 (1984)
86. Chuang, H. Y. K., Sharpton, T. R. and Mohammad, S. F.: Int. J. Artificial Organs *6*, 199–206 (1983)
87. Chuang, H. Y. K. and Andrade, J. D.: J. Biomed. Materials Res. *19*, 813, 1985
88. Furie, B., Blanchard, R. A., Robinson, D. J., Tai, M. M., and Furie, B. C.: Methods Enzym., Langone, J. J. and Vunakis, H. V. (eds.), New York: Academic Press, 60–82 (1982)
89. Grinnell, F., Feld, M. K.: J. Biol. Chem. *257*, 4888–4893 (1982)
90. Uzgiris, E. E.: J. Cellular Biochem., Suppl. 9B, 138 (Abst. 766), 1985
91. Walton, A. G. and Koltisko, B.: Biomaterials, Adv. Chem. Series #199, Cooper, S. L. and Peppas, N. A. (eds.), Washington DC: American Chemical Society, 245–264 (1982)
92. Hlady, V., Van Wagenen, R. A., and Andrade, J. D.: Protein adsorption, vol. 2, chap. 2, Andrade, J. D. (ed.), New York: Plenum Press, 81–120, 1985
93. Chan, B. M. C. and Brash, J. L.: J. Colloid Interface Sci. *82*, 217–225 (1981)
94. Brynda, E., Houska, M., Pokorna, Z., Cepalova, N. A., Moiseev, Y. V., and Kalal, J.: J. Bioeng. *2*, 411–418 (1978)
95. Jennissen, H. P.: J. Chromatog. *159*, 71–83 (1978)
96. Jennissen, H. P.: Z. Physiol. Chem. *357*, 1727–1733 (1976)
97. Jennissen, H. P. and Botzet, G.: Int. J. Biologic Macromolecules *1*, 171–179 (1979)
98. Vroman, L. and Adams, A. L.: J. Colloid Interface Sci., in press (1986)

99. Schwick, H.-G. and Haupt, H.: Angew. Chem. Int. Ed. *19*, 87–99 (1980)
100. Ash, S. G.: Colloid Science *1*, Everett, D. H. (ed.), London: The Chemical Society, 103–122 (1973)
101. Adams, A. L., Fischer, G. C., Munoz, P. C., Vroman, L.: J. Biomed. Materials Res. *18*, 643–654 (1984)
102. Horbett, T. A.: Thromb. Haemostas *51*, 174–181 (1984)
103. Brash, J. L. and ten Hove, P.: Thromb. Haemostas *51*, 326–330 (1984)
104. Breemhar, W., Brinkman, E., Ellens, D. J., Beugeling, T., Bantjes, A.: Biomaterials *5*, 269–274 (1984)
105. Lovelock, J. E. and Porterfield, J. G.: Nature *167*, 39–41 (1951)
106. Sawyer, P. N.: Ann. N. Y. Acad. Sci. *46*, 561 (1983)
107. Zisman, W. A.: Contact Angle, Wettability, and Adhesion, Fowkes, F. M. (ed.), Adv. Chem. Ser. *43*, 1–51 (1964)
108. Baier, R. E.: Bull. N. Y. Acad. Med. *48*, 257–272 (1972)
109. Fowkes, F. M.: Chemistry and physics of interfaces, II, Ross, S. (ed.), Washington DC: Amer. Chem. Soc. 154–167 (1971)
110. Good, R. J. and Elbing, E.: Chemistry and physics of interfaces, II, Ross, S. (ed.), Washington DC: Amer. Chem. Soc. 71–96 (1971)
111. Lyman, D. J., Muir, W. M., and Lee, I. J.: Trans. Amer. Soc. Artif. Int. Organs *11*, 301–317 (1965)
112. Andrade, J. D.: Med. Inst. *7*, 110–120 (1976)
113. Wichterle, O. and Lim, D.: Nature *185*, 117 (1960)
114. Andrade, J. D., King, R. N., Gregonis, D. E., and Coleman, D. L.: J. Polymer Sci., Polymer Symp. *66*, 313–336 (1979)
115. Anderson, J. M. and Koftke-Marchant, K.: Crit. Rev. Biocomp. *1*, 111–204 (1985)
116. Salzman, E. W., Lindon, J., Brier, D., Merrill, E. W.: Ann. N. Y. Acad. Sci. *283*, 114–124 (1977)
117. Kim, S. W. and Lyman, D. J.: Applied Polymer Symposium *22*, 289–297 (1973)
118. Chang, T. M. S.: Canadian J. Physiology and Pharmacology *52*, 275–285 (1974)
119. Andrade, J. D., Coleman, D. L., Kim, S. W., and Lentz, D. J.: Artificial Liver Support, Williams, R., and Murray-Lyon, I. M. (eds.), Pitman Medical, 84–93 (1975)
120. Sato, T. and Ruch, R.: Stabilization of colloidal dispersions by polymer adsorption, New York: Dekker 1980
121. Merrill, E. W. and Salzman, E. W.: ASAIO J. *6*, 60–64 (1983)
122. Gregonis, D. E., Buerger, D. E., Van Wagenen, R. A., Hunter, S. K., Andrade, J. D.: Biomaterials 84, Trans. Soc. Biomaterials *7*, 766 (1984)
123. Mori, Y. and Nagaoka, S.: Trans. Am. Soc. Artif. Internal Organs *28*, 459–463 (1982)
124. Bell, G. I., Dembo, M., Bongrand, P.: Biophys. J. *45*, 1051–1064 (1984)
125. Chenoweth, D. E.: Artificial Organs *8*, 281–287 (1984)
126. Sevastianov, V. I. and Tseytlina, E. A.: J. Biomed. Materials Res. *18*, 969–978 (1984)
127. Matsuda, T., Takana, H., Hayashi, K., Taenaka, Y., Takaichi, S., Umezu, M., Nakamura, T., Iwata, H., Nakatani, T., Tanaka, T., Takatani, S., and Akutsu, T.: Trans. Am. Soc. Artif. Intern. Organs. *30*, 353–358 (1984)
128. Moacanin, J. and Kaelble, D.: Polymer *18*, 475 (1977)
129. Hoffman, A. S.: Biomaterials, Adv. Chem. Series #199, Cooper, S. L., and Peppas, N. A. (eds.), Washington DC: American Chemical Society, 1–8 (1982)
130. Baier, R. E.: Artificial Organs *2*, 422–426 (1978)
131. Williams, D. F. and Bagnall, R. D.: Fundamental Aspects of Biocompatibility *2*, Williams, D. F. (ed.), CRC Press, 113–127 (1981)
132. Chiu, T.-H., Nyilas, E., and Turcotte, L. R.: Trans. Am. Soc. Artif. Intern. Organs *24*, 389–402 (1978)
133. Nagaoka, S., Mori, Y., Takiuchi, H., Yokota, K., Tanzawa, H., and Nishiumi, S.: Polymers as Biomaterials, Shalaby, S. W., Hoffman, A. S., Ratner, B. D., and Horbett, T. A. (eds.), Plenum Press, 361–374 (1984)
134. Van der Scheer, A., Feijen, J., Elhorst, J. K., Dagneaux, G. K., and Smolders, C. A.: J. Colloid Interface Sci. *66*, 136–145 (1978)
135. Jennissen, H. P.: Protein adsorption, vol. 2, chap. 9., Andrade, J. D. (ed.), New York: Plenum Press, 295–320 (1985)

136. Brash, J. L., Uniyal, S., and Samak, Q.: Trans. Am. Soc. Artif. Intern. Organs *20*, 69 (1974)
137. Thompson, N. L., Burghardt, T. P., and Axelrod, D.: Biophys. J. *33*, 435 (1981)
138. Norde, W.: Commun. Agric. Univ. Wageningen *76-6* (1976)
139. Hlady, V., Furedi-Milhofer, H.: J. Colloid. Interface Sci. *69*, 460 (1979)
140. Fair, D. B. and Jamieson, A. M.: J. Colloid. Interface Sci. *77*, 525 (1980)
141. Norde, W.: Protein adsorption, vol. 2, chap. 8, Andrade, J. D. (ed.), New York: Plenum Press, 263–294 (1985)
142. Norde, W. and Lyklema, J.: Protein adsorption, vol. 2, chap. 7, Andrade, J. D. (ed.), New York: Plenum Press, 241–262, 1985
143. Silberberg, A.: Protein adsorption, vol. 2, chap. 10, Andrade, J. D. (ed.), New York: Plenum Press, 321–338, 1985
144. Koltisco, B. and Walton, A. G.: Protein adsorption, vol. 2, chap. 7, Andrade, J. D. (ed.), New York: Plenum Press, 217–240, 1985
145. Harrick, N. J.: Internal reflection spectroscopy, New York: Wiley, 1967
146. Hansen, W. N.: J. Opt. Soc. Amer. *58*, 380 (1968)
147. Hansen, W. N.: Adv. Electrochem. Electrochem. Eng. *9*, Muller, R. H. (ed.), 1 (1973)
148. Born, M. and Wolf, E.: Principles of optics, 5th ed., New York: Pergamon Press, 1975
149. Hirschfeld, T.: Can. Spectrosc. *10*, 128 (1965)
150. Harrick, N. J. and Loeb, G. I.: Anal. Chem. *45*, 687 (1973)
151. Kronick, M. N. and Little, W. A.: J. Immunol. Methods *8*, 235 (1975)
152. Hirschfeld, T. and Block, M. J.: Opt. Eng. *16*, 406 (1977)
153. Hirschfeld, T., Block, M. J., and Mueller, W.: J. Histochem. Cytochem. *25*, 719 (1977)
154. Rockhold, S. A., Quinn, R. D., Van Wagenen, R. A., Andrade, J. D., and Reichert, M.: J. Electroanal. Chem. *150*, 261 (1983)
155. Burghardt, T. P. and Axelrod, D.: Biophys. J. *33*, 455 (1981)
156. Rainbow, M. R., Atherton, S., and Eberhart, R. C.: J. Biomed. Mater. Res., in press, 1986
157. Thompson, N. L. and Axelrod, D.: Biophys. J. *43*, 103 (1983)
158. Burghardt, T. P. and Axelrod, E.: Biochemistry *22*, 979 (1983)
159. Hlady, V., Reinecke, D. R., and Andrade, J. D.: J. Colloid Interface Sci., in press, 1986
160. Newby, K. N.: M.S. thesis, University of Utah, Salt Lake City, Utah, U.S.A., 1984
161. Suci, P.: M.S. thesis, University of Utah, Salt Lake City, Utah, U.S.A., 1984
162. Cornell, D. G.: J. Colloid Interface Sci. *98*, 283 (1984)
163. Payser, P. and Stromberg, R. R.: J. Phys. Chem. *71*, 2066–2074 (1967)
164. Gendreau, R. M., Leininger, R. I., Winters, S., and Jakobsen, R. J.: Adv. Chem. Ser. *199*, 371 (1982)
165. Gendreau, R. M., Winters, S., Leininger, R. I., Fink, D., Hassler, C., and Jakobsen, R. J.: Appl. Spectrosc. *35*, 353 (1981); also, Gendreau, R. M., Ibid., *36*, 47 (1982)
166. Bellissimo, J. A. and Cooper, S. L.: Trans. Am. Soc. Artif. Intern. Organs *30*, 359 (1984)
167. Castillo, E. J., Koenig, J. L., Anderson, J. M., and Lo, J.: Biomaterials *5*, 319 (1984)
168. Fink, D. J. and Gendreau, R. M.: Anal Biochem. *139*, 140 (1984)
169. Reichert, W. M. and Andrade, J. D.: Surface and Interfacial aspects of Biomedical Polymers, vol. 1, chap. 12, Andrade, J. D. (ed.), New York: Plenum Press, 421–442, 1985
170. Iwamoto, R., Ohta, K., Miua, M., and Mima, S.: Appl. Spectroscopy *36*, 584 (1981)
171. Chen, Y. J., Chen, W. P., and Burstein, E.: Phys. Rev. Lett. *36*, 1207 (1976)
172. Cotton, T. M.: Protein adsorption, vol. 2, chap. 4, Andrade, J. D. (ed.), New York: Plenum Press, 161–188, 1985
173. Aurengo, A., Masson, M., Duyperat, R., Levy, Y., Hasmonay, H., and Barbillat, J.: Biochem. Biophys. Res. Commun. *89*, 559 (1979)
174. Cotton, T. M., Schultz, S. G., and Van Duyne, R. P.: J. Am. Chem. Soc. *102*, 7960 (1980)
175. Trurnit, H. J.: Arch. Bioch. Bioph. *47*, 251 (1953)
176. Trurnit, H. J.: Arch. Bioch. Bioph. *51*, 176 (1954)
177. Poste, G. and Moss, C.: Prog. Surface Sci. *2*, 139–232 (1972)
178. Morissey, B. W., Smith, L. C., Stromberg, R. R., and Fenstermarker, C. A.: J. Colloid Interface Sci. *56*, 557 (1976)
179. Jonsson, U., Ivarsson, B., Lundstrom, I., and Berghem, L.: J. Colloid Interface Sci. *90*, 148 (1982)
180. Horisberger, M.: Biochem. Biophys. Acta *632*, 298 (1980)
181. Elwing, H. and Stenberg, M.: J. Immunol. Methods *44*, 343 (1981)

182. Elwing, H., Dahlgren, C., Harison, R., and Lundström, I.: J. Immunol. Methods *71*, 185 (1984)
183. Mandenius, C. F., Welin, S., Danielsson, B., Lundström, I., and Mosbach, K.: Anal. Biochem. *137*, 106 (1984)
184. Cuypers, P. A., Hermens, W. Th., and Hemker, H. C.: Anal. Biochem. (84), 56 (1978)
185. Cuypers, P. A., Corsel, J. W., Janssen, M. P., Kop, J. M. M., Hermens, W. Th., Hemker, H. C.: J. Biol. Chem. *258*, 2426 (1983)
186. Cuypers, P. A.: Ph.D. thesis, University Limburg, Maastricht, The Netherlands, 1976
187. Cuypers, P. A., Corsel, J. W., Janssen, M. P., Kop, J. M. M., Hermens, W. Th., Hemker, H. C.: Surfactants in Solution, vol. 2, Mittal, K. L. and Lindman, B. (eds.), New York: Plenum Press, 1301 (1984)
188. Proceedings of the conference on protein and polyelectrolyte adsorption, Potsdam, N.Y., June, 1985: J. Colloid Interface Sci., in press (1986)
189. Eisenberg, D., Schwarz, E., Komaromy, M., Wall, R.: J. Molec. Biol. *179*, 125–142 (1984)
190. Rose, G. D., Gierasch, L. M., Smith, J. A.: Adv. Prot. Chem. *37*, (1985)
191. Matthew, J. B.: Ann. Rev. Biophysics Biophys. Chem. *14*, 387–418 (1985)
192. Vroman, L., Adams, A. L., Klings, M., Fischer, G.: in Baier, R. E. ed., *Applied Chemistry at Protein Interfaces*, Amer. Chem. Soc., Adv. Chem. Series #145, Washington D.C. (1975) 255–289
193. Reichert, W. M., Newby, K., Andrade, J. D.: Trans. 11th Ann. Meet. Soc. for Biomaterials *8* (1985) 5

Editor: K. Dušek
Received September 24, 1985

Medical Polymer Adhesives

T. E. Lipatova
Institute of Organic Chemistry,
Academy of Sciences, Ukrainian SSR, Kiev/USSR

A critical survey of literature data on medical polymer adhesives is presented. Various classes of polymers are discussed from the point of view of their application in surgery.

Cyanacrylate adhesives are described along with some data on the strength of adhesive joints of living tissues, the processes of biodegradation and elimination from the organism, and the reaction of living tissues to the adhesives. Among other classes of medical adhesives, epoxy compounds and gelatin-resorcin-formaldehyde resins and their applications are discussed (e.g., experimental glueing of bones with epoxy resins).

Special attention is given to the polyurethane adhesives. A generalized approach to the creating of medical glues is proposed based on the author's experimental results. This approach includes consideration of phenomena at the living tissue-adhesive interface, analysis of weak border layer formation, and analysis of regeneration processes by adhesive contact with the living tissue. The general principles of medical biocompatible polymer synthesis have been formulated and some ways of solving arising problems are proposed.

Advances in Polymer Science 79

Introduction

Polymer adhesives have been used for about 30 years in medical practice for joining different tissues of the organism. The process of glueing and the effect of the adhesive on the organism have been studied mainly by specialists in medicine and biology, focussing on the tissues surrounding the adhesive joint as well as on the organism itself. Such investigations considered, e.g., the elimination of adhesive fragments from the organism by means of ^{14}C labeling. Extremely few joint physico-chemical and biological investigations have been carried out till the present time. The problem of glueing living tissues is quite complicated for various reasons: physico-chemical aspects (phenomena at the adhesive-living tissue interface, structure of the adhesion joint, the characteristics of its destruction, changes in its structure, etc.), biological aspects (general and local toxicity, regenerative processes at the boundary of the adhesion joint, etc.), and finally, and very importantly, the mechanical properties of the adhesion joint. It is not only the adhesive characteristics of the adhesion joint that are important in this aspect but also the cohesion strength of the joint. All these aspects should be studied simultaneously and the results compared. This is the only scientific approach to develop medical adhesives that are required for different tissues and under various surgical conditions.

The present article does not attempt to cover the entire range of existing adhesives and experimental investigations of adhesion joints within the last 30 years. Rather, the article focuses on adhesives that have been studied experimentally, more or less in detail, which are used in medical practice. It should be noted that maximum interest to the problem of glueing tissues was recorded in the 60s and the early 70s; the interest then diminished greatly, although the significance of the problem has not decreased.

A cured adhesive is a peculiar endoprothesis in the organism; the difference to common endoprotheses is its formation under the effect of the living tissue surface. The surface of body exerts a comprehensive effect on the formation of polymers and their properties [1]. The effect of a living tissue on the structure of the forming adhesion joint has not been investigated to date. An adhesive endoprothesis, according to our knowledge [2]), is an endoprothesis of temporary action, i.e., having performed its function in the organism, it eventually is degraded. Its fragments must then be eliminated from the organism. Certain endoprotheses are known to have lifelong action, the time of degradation in the organism greatly exceeds the life of the patient. It is well known today [3,4] that all polymeric materials are destroyed in the organism.

What are the requirements of modern medicine to an adhesive?

I. The adhesive should tightly hold the edges of a moist operative wound.

II. The adhesive should not have any type of or local toxic effect on the organism and surrounding tissues.

III. The adhesive, its initial components and products of biodegradation should not have blastomogenic effects.

IV. The adhesive should not be allergenic.

V. The adhesion joint should not prevent (ideally it should contribute to) the formation of connective tissue scars and regeneration of the tissues.

VI. The products of biodegradation of the adhesive should be eliminated from the organism by common means, without cumulating in organs or tissues.

VII. The physico-chemical and mechanical properties of the adhesion joint should correspond to the given surgical situation wherein the adhesive is used.

We have considered it necessary to add one more requirement to the adhesive: it should glue under conditions of a moist operative area, because intensive drying of the tissues prior to glueing would not contribute to the normal occurence of regenerative processes. The manner in which this is achieved will be discussed below.

One more circumstance should be borne in mind as it refers equally to adhesives and endoprotheses: they should meet the criterion of medical purity, i.e., they should not contain any admixtures that are harmful to the organism and can diffuse from the material, while the polymers may contain harmless admixtures, e.g., certain plasticizers and stabilizers.

1 Cyanoacrylate Adhesives

Cyanoacrylate adhesives are the most widespread type of adhesive that are applied today for glueing traumatized tissues. They have been subjected to long-term improvement by introducing different alkyl substituents and even fluorine atoms [5,6].

The class of cyanoacrylate adhesives may be expressed by the following formula

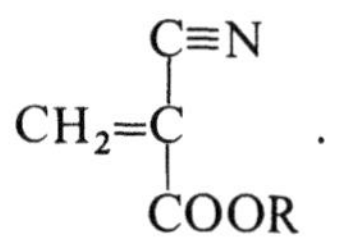

Macromolecules are formed by opening the double bond

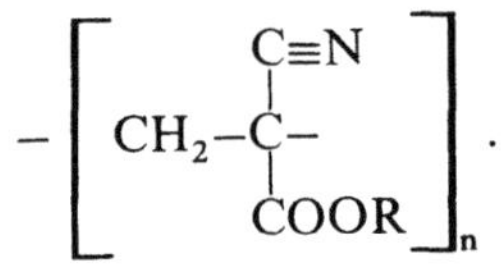

A group of adhesives (Estman-910) has been developed on the basis of methyl-α-cyanoacrylates [7,8], adhesives of the type Alpha S(AAS) and Aron Alpha $S_2(AAS_2)$ on ethyl-α-cyanoacrylate [9]. Butyl and isobutyl homologues of Estman-910 have served as a basis for developing the adhesives Histoacryl blau and Histocol TIOOB, respectively [10,11]. Adhesives have been developed and tested on the basis of hexyl- and decylcyanoacrylates [12], as well as fluorocyanoacrylate adhesives [5,6]. Adhesives on the basis of copolymers of different cyanoacrylates have also been tested [13]. Cyanoacrylate adhesives have been modified by introducing a plasticizer. For example, dibutyl phthalate has been used in adhesive M—I on the basis of ethyl-α-cyanoacrylate [14], while adhesive M-3 on the same basis and with the same plasticizer contained also a thickening agent, namely, polyethyl-α-cyanoacrylate (Ref. [15], p. 28). Poly(vinyl acetate) was added to adhesive MK-2 and, according to the authors, it acts as a thickening agent and a plasticizer [16]. Compositions on the basis of cyanoacrylate adhesives are supplemented with dyes (to improve the conditions of observing the adhesion joint) and drugs (for example, Shostakovsky balsam) in adhesive MK-6 [17].

It is clear from this brief list of adhesives on the basis of alkyl-α-cyanocrylates that these adhesive have been modified by replacing the alkyl substitute in the acryl group as well as by introducing various additives characterized by plasticizing or medicinal properties.

Despite the fact that cyanoacrylate adhesives have been studied in great detail, there is no clear idea about the way adhesion of the compound to living tissues is effected. Most thorough drying of the operative area is required in case of applying a cyanoacrylate adhesive, as it is evident that moisture is not displaced by the adhesive from the tissue surface. It is supposed [18] that cyanoacrylates are polymerized by an anionic mechanism. Consequently, the proton donor compounds terminate the process. Hence, again most thorough drying of the wound before glueing is significantly important. It follows also that the adhesion properties of the adhesive are determined by the presence of polar CN and ester groups that interact with amino and carboxyl groups of protein molecules of tissues freed of moisture.

1.1 Strength of Glueing with Cyanoacrylate Adhesives

The adhesion joint should feature satisfactory strength and tightness. It should be asked what parameters are suitable for estimating the strength of glueing before using the adhesive in a clinic.

A common test is not possible because the strength of glueing depends on the nature of the tissue, conditions of applying the adhesive, its amount, and kind of clinical pathology.

Many methods have been proposed to estimate the strength of adhesion joints. We shall deal only with methods by which the strength of gluieng living tissue has been investigated, because all the methods proposed for testing medical adhesives by glueing together different materials, even natural ones (e.g., catgut), are not at the least indicative for a real situation.

Heis et al. [19] joined superimposed skin incisions by means of the Estman-910 adhesive in experiments on rats. Having cut out a strip of glued skin of a certain size, one of its ends was immobilized, while a containement was suspended from the other end. The containement was gradually filled with water which enabled to determine the breaking load and compare it with the load that destroyed sutured skin. It is impossible, however, to estimate the strength of glueing in this case though the authors came to the conclusion that an adhesion joint was not inferior to a sutured one.

Woodward et al. [20] also investigated the shear strength of animal skin sections glued with a cyanoacrylate adhesive and cut off by a dermatome. Unfortunately, their method is complicated and it is impossible to perform a comparative estimation of the strength of glueing with different cyanoacrylates.

A method for estimating the strength of an adhesion joint under conditions approaching real ones has been proposed [21]. When defects in the aortal wall of dogs were closed by glueing on different alloplastic materials, the arterial pressure was raised in the animals to 300–400 mm Hg by injecting adrenaline. Disturbance in thightness was effected by breaking the adhesion joint at a pressure of about 350 mm Hg, which is a high index of strength. The same method has been used in other studies [22] investigating the effectiveness of closing kidney wounds by glueing on strips of amnion. In our opinion, the maximum approximation to a real surgical situation has been attained in the latter work; therefore, these results are of definite interest for clinical surgery.

The method of investigating the strength of an adhesion joint for plastic surgery on the front abdominal wall experimentally performed on rabbits [23], has attracted some attention. The operation consisted in making an incision in the skin and muscular aponeurotic layer in the front abdominal wall over a length of 5–6 cm. The peritoneum was carefully separated. Plastic surgery of the abdominal wall was performed by the method of Sapezhko [23]. When the adhesive was applied on the surface of one part of the aponeurotic tissue, the latter was overlapped by another part of the tissue. The joint was held during polymerization of the adhesive by two provisory π-shaped ligatures which were removed 2–3 minutes after glueing. The glued sections of the front abdominal wall in the form of plates of correct rectangular shape were dissected at different periods of time after the operation, which made it possible to determine the area of the adhesion joint with relative accuracy. The peripheral ends of the samples were fastened to duralumin plates that were clamped in the terminals of an automatic tearing machine (ZM-4). The chosen model was very close to a real situation. Adhesion of homogeneous tissues made it possible to estimate the adhesion strength correctly. Moreover, the adhesion joint was permanently under a load owing to tensioning of the aponeurotic strips during the operation and pressure of the internal organs on the front abdominal wall. This model was suitable for determining the integrity of the adhesion joint when palpating the front abdominal wall of the animals. A comparison of the results of the adhesion strength of two chemically entirely different adhesives by means of the above model will be discussed below.

From the results on model systems it follows that a model should approach a real surgical situation as much as possible. For example, only having tested an adhesive on the skin does not allow for applying these results to other tissues.

1.2 Biodegradation and Elimination from the Organism

The adhesion joint should be replaced by regenerated tissue; therefore, the polymeric material should be destroyed, then metabolized and eliminated from the organism. Elimination of the products of biodegradation was studied [24–26] by means of ^{14}C labelina in different groups of the cyanoacrylate adhesive Cyacryn (nitrile, β-atom of the main chain and in the ethoxy group). It had been established that elimination of these products from the organism occurred in the urine, feces, and expired air. This process started several minutes after glueing. The products of Cyacryn degradation cumulated partially in different organs for periods up to 10 months. Maximum accumulation of the products of biodegradation occured in the pituitary, brain, thyroid gland, and liver. The authors had good reason to believe that the Cyacryn polymer was subject to hydrolysis in the animal organism by the $COOC_2H_5$ groups. Cleavage of the CN group with the formation of ammonia also occurred.

Degradation of the main (carbon) chain by oxidation to CO_2, water, and the formation of formaldehyde occurred at a later time. The idea of a cellular mechanism of cyanoacrylate adhesive (Estman-910) resorption has been rejected [27]. After applying an adhesive to different regions of an organism, it was found that the rate of adhesive degradation varied and, therefore, detoxification of the products and their degradation occurred at different periods of time, owing to which the toxicity of the adhesive was unequal in different cases [28].
However, it should be noted that we have not found in the literature any record of investigations on the degradation of cyanoacrylate adhesives in model media nor any comparison of such results with observations made in vivo. All the assumptions about the degradation of cyanoacrylate adhesives are rather convincing but they have not been confirmed. It is obvious that a change in the nature of the substituent in the oxyalkyl group of cyanoacrylates will cause a change in the rate of polymerization and in the rate of its degradation [29,30]. Direct quantitative comparison of this issue has not been made yet.

1.3 Reaction of Tissues

When considering the reaction of tissues to the introduction of an adhesive, it was interesting to establish whether the length and structure of the alkyl substituent in the alkoxyl group of cyanoacrylate render any effect on the toxicity of the monomer. The degree of toxicity of various cyanoacrylate monomers was estimated in Ref. [31] by the tissue reaction following the implanting of ivalon sponges impregnated with these monomers, in white rats. The tissue reaction was estimated by the proliferation of fibroblasts and the production of collagen fibers within the sponges. The authors considered also the tissue reaction to each preparation individually. It was established that monomers of hexyl and decyl cyanoacrylates had not retarded the formation of collagen and reproduction of fibroblasts, and caused less tissue necrosis as compared with methyl-2-cyanoacrylate and Estman-910. All investigated cyanoacrylates appear to have polymerized in the presence of a large interface area of the sponge. The effect of the surface area was reflected in the values of the molecular mass of the produced polymers, therefore it was incorrect to compare the effect of the polymer produced in the presence vs. absence of the sponge surface. The conditions used by the authors for different cyanoacrylates in the presence of a sponge referred evidently to polymers and not to monomers. Indeed, it had been shown [29,30] that the longer the alkoxyl group in the alkoxyl substitute and the faster the cyanoacrylate polymerized, the lower was its toxicity and the slower its degradation in the organism. Attention should be paid to the fact that polymers produced from one and the same cyanoacrylate but differing in molecular mass have different toxic effects [32,33]. Implantation of an ethyl acrylate polymer with a high molecular mass caused tissue necrosis with considerable granulomatosis.

A good method, enabling to estimate the histotoxicity of the adhesive and separate its action from the result of traumatizing the organ, consists in applying the adhesive on uninjured parts of the organ or tissue. The application of 0.002 ml of methyl-2-cyanoacrylate to liver, subcutaneous fat, or bone tissue causes considerable necrosis and suppuration [20]. Its homologues (hexyl- and decyl-2-cyanoacrylate) cause a more favorable reaction in agreement with other data [31]). The decrease of histotoxicity is associated with a reduction of the relative content of CN groups in the material [34].

Evidently, the lower rate of degradation of this polymer, which is associated with the secretion of toxic compounds, is also of certain significance [29,30].

Various cyanoacrylate adhesives have been investigated experimentally by applying the latter to uninjured surfaces of the large intestine, decapsulated kidney, walls of vessels, etc. We shall not dwell in detail on the results of these investigations, firstly, because they are incomparable owing to the fact that different tissues were investigated, and, secondly, because they were obtained under different conditions. We consider it necessary to discuss in greater detail results obtained on the adhesive Estman-910 [35], which is used widely in clinical work. The adhesive was spread in the form of an application on the surface of the brain and greater somatic arteries of rabbits and cats. The adhesive films adhering to the cerebral cortex, caused gliosis, cellular infiltration, and death of neurons at a depth of 2–3 mm. Erosion was observed in the adventitia and muscular layer of the arteries. It is evident from this example that adhesives should be used selectively for different tissues. Polyurethane adhesive KL-3, which will be discussed in a separate part of this article, proved harmless in contact with brain. The disadvantage of all cyanoacrylate adhesives is their monolithic character; if is frequently a rigid film having an unfavorable effect on the tissue. The highly sensitive tissue of the brain reacts to these factors more intensely than other tissues.

General stutements on the experience in using cyanoacrylate adhesives stated 15 years ago [36] seem to still apply today. According to these authors [36], butyl and isobutyl cyanoacrylate adhesives are less toxic than methyl cyanoacrylates. The disadvantage of these adhesives is their slow biodegradation which, however, presents no obstacle for the healing processes and causes no blastomogenic degeneration of the tissues. There have been many publications on the application of cyanoacrylate adhesives in clinical work in different areas of surgery. These results have been comprehensively summarized [37].

It should be noted in conclusion that cyanoacrylate adhesives, despite their widespread application in clinical work, are still being investigated experimentally. The investigations are of purely biological character. Investigations of cyanoacrylate adhesives in the physico-chemical sense with respect to the living tissue-adhesive interface have not been carried out yet. The growing interest in the research on new types of adhesives is apparently the reason for this shortcoming.

2 Other Types of Medical Adhesives

Publications on other classes of medical adhesives are rather scarce and incomplete. Past investigations mainly concerned biological aspect, while chemical relationships are treated rather empirically and are reduced to finding optimum mixing ratios.

Gelatin-resorcin-formaldehyde adhesives (GRF) have been studied [38–40]. These adhesives compared with cyanoacrylates have an advantage in that they join tissues reliably under conditions of a moist operative area and undergo rapid degradation in the organism. The rate of curing is adjusted easily by changing the ratio of components.

The intensity of the inflammatory reaction when using GRF adhesives is diminished easily by reducing the content of formaldehyde and by decreasing the pH approximately to 3 [40]. A comparative investigation of closing liver wounds in dogs by means of GRF adhesives and butyl and isobutyl cyanoacrylate adhesives revealed a considerable inflammatory reaction for the GRF adhesive [41] which, however, was subjected to a faster biodegradation in the organism. Contrary to these results, it was found that the GRF adhesive in glued liver wounds interfered in regular healing processes owing to the irritating effect of formaldehyde. The chemical mechanism of the GRF adhesive degradation has not been investigated, the changes in the physico-chemical characteristics of the adhesive joint are unknown, and even the gluieng strength has not been determined.

Epoxy adhesives were quite popular at one time. They exert high strength and are used, or, to be more exact, were used for glueing bone tissues. The inconvenience and difficulty of using these adhesives consists in the fact that even a minor amount of moisture prevents the curing of the adhesive [44]. The epoxy adhesive "osteoplast" was used in clinical work for glueing bones [45]. Application, the bone surface freed of the periosteum has to be dried thoroughly by a mixture of ethanol and ethylether (1:1) and dried with a fan. It can be assumed that this treatment did not contribute to the healing processes. Information on the action of epoxy adhesives and their compositions, e.g., with cyanoacrylate adhesives [46] for gluing tissues, is quite contradictory and doubtful.

We have discussed the major classes of adhesive compounds which have been widely investigated experimentally and used in clinical work. It follows that the investigations were mainly of biological character and no attemps have been made to analyze the glueing process from the physico-chemical point of view or to compare the results with the biological characteristics of the adhesives. Such an attempt has been made by us with the polyurethane adhesive KL-3 [47], which is currently widely used in surgery. Before presenting our ideas on glueing living tissues and on the approaches to the formulation of adhesives, we shall first consider polyurethane adhesives. It should be noted that this type of adhesives occupies the second place after the cyanoacrylates.

3 Polyurethane Adhesives

Polyurethane prepolymer was used for the fist time for glueing tissues in 1958 [48]. The fundamental notions of adhesion and glueing employed later in developing a large number of technical adhesives had not been elaborated at that time in sufficient detail.

Physico-chemical approaches have found widespread application for the development of adhesives. These approaches are based on adjusting the conditions of wetting the surfaces to be glued by liquid polymer adhesives, adjusting internal stress in adhesive joints, and modifying adhesives by compounds capable of forming a chemical bond with the base of the adhesive as well as with the surface being glued [49,50].

According to modern ideas, the basic condition for a good adhesive joint consists in appreciable wetting of the surface with the adhesive and spreading of the latter, as well as high energy of cohesion of the material in the zone of contact with the surfaces being glued. For this purpose, it is necessary to adjust the surface energies of the glued surfaces and the adhesive. If the surface energy is low, the adhesive is not spread and adhesion is not sufficient. The latter problem becomes very important in glueing living tissues, whose superficial characteristics are not taken into consideration (in particular, owing to the lack of reliable methods of evaluating these properties). The surface-active properties of the adhesive and their changes during the curing process can be affected by adjusting the properties of polymer adhesives by introducing reactive surfactants [50], but this approach has not yet been applied to the development of medical adhesives.

Polyurethane adhesives (bone-glue, polyurethane foam, Ostamer) were first used for joining bones [51,52]. The adhesive "Ostamer" was composed of a prepolymer and an accelerator which were mixed before use at a ratio of 4:1. Curing occurred in 25–30 minutes. Evidently, this curing time is unacceptable under clinical conditions. The adhesive reached maximum strength in 1–2 days. The experimental and clinical results of those years in glueing bones should be considered unsatisfactory [53-55]. When glueing soft tissues, a polyurethane prepolymer was used with success for tightening joints on transverse and longitudinal incisions on the aorta, inferior femoral vein, and femoral artery in dogs [56]. The authors associate the positive reults with the advantages of the adhesive (which we consider unfounded; e.g., the adhesive was chemically inert), and with the comprehensive adhesion of the polymer to the wall of the blood vessel. The reasons and mechanism of adhesion of polyurethane to living tissue has not been comprehensively investigated. The assumption was expressed that free NCO groups of the prepolymer were chemically bound to molecules of living tissues containing functional groups with mobile hydrogen [56] (cf. also Ref. [57]. A prepolymer on the basis of poly(oxyethylene) glycol or poly(oxypropylene) glycol and toluylene diisocyanate (TDI) was used with success for glueing muscles in cattle [58]. This adhesive contained the same chemical components as adhesive KL-3 (discussed below).

Experimentally, a combination of TDI with castor oil made it possible to ensure good adhesion to soft tissues, but the curing time (10 minutes) did not satisfy surgical requirements [59].

An adhesive prepared on the basis of active 6-chloro-2,4,5-trifluoro-1,3-phenylene diisocyanate, castor oil, pyridine, and trace amounts of other additives indicated negligibly minor adhesion to living tissues [60]. The results [48] indicate clearly, that chemical interactions have no definite significance in adhesion to living tissues. Indeed, the structure of the surface of any living tissue is quite complicated; the surface is composed of microsections capable of hydrophobic and hydrophilic interactions. This has been confirmed [61] by an investigation of the heart intima and blood vessels using the Zisman method. The contact wetting angles were determined by using liquids with different surface tensions. It was found that the investigated surfaces were composed of hydrophilic and hydrophobic sections. The critical surface tension of hydrophobic and hydrophilic sections was 29 dyne/cm and 68.5 dyne/cm, respectivelly.

The question arises: what sections of living tissue should the development of medical adhesives be based on? According to our opinion, the optimum adhesion is attained with surfaces characterized by high surface energy. Hence, it is necessary to consider hydrophilic sections. These are usually coated with a layer of tissue fluid, which should be removed by substitution for a surface-active adhesive, or by chemical bonding, or both. The second variant has been performed by isocyanate [48]. Additional investigations are necessary to determine the effect of added pyridine or tetrahydrofurane. In some cases excessive diisocyanate is necessary in the prepolymer mixture to ensure good adhesion to biological tissues [60,62].

4 Possibilities of Improving the Adhesion of Adhesives to Living Tissues

Experimental results on polyurethane adhesives suggested the idea of improving the adhesion of polymer adhesives to living tissues by means of certain monofuctional additives to the adhesive, or by using them for pretreatment of the living tissue surface [63]. Adhesion was evaluated on the basis of pressure measured by a pneumatic tensiometer [60]. The pressure value was recorded instead of the tension value in the adhesion joint. Investigations were carried out on tissues dissected from animals one hour after curing of the adhesive. A series of commercial polymers were tested in the experiment. No polymer exhibited adhesion to living tissues without the addition of benzoyl chloride. When 10% benzoyl chloride was added, adhesion of all the tested adhesives (nitrile rubber, silicone elastomers, and polystyrene) to living tissues increased considerably. It was also established that adhesion did not depend on whether the living tissues were pretreated with benzoyl chloride in methylene chloride or whether the solution was added to the adhesive. An interesting experimental result of these investigations is the fact that the molecular mass had no effect on the adhesion to living tissues. The authors in some cases used polymers in solution and evidently considered the time of solvent (methylene chloride) volatilization as the curing time. Nitrile rubber (trademark Pliobond) was used for investigating the effect of benzoyl chloride, added to the adhesive, on adhesion. Adhesion did not occur if the stated compound was not added. The adhesion increased linearly upon addition of 1 to 3% benzoyl chloride. Adhesion remained constant beyond the 3% limit. The authors considered that this value of adhesion (it would be more correct to use the term "strength") resulted from the cohesion damage by the living tissue and not by the alluation of the adhesion joint. By means of a pneumatic tensiometer, it was possible to observe the connective tissue fibers separating the living tissue and penetrating the surface of the polymer layer. Similar results were obtained when substituting benzoyl chloride for *para- and meta-benzoyl chloride*, which indicates that the position of the substituent in the benzoyl chloride molecule has no significance. The results obtained by the authors of this review are presented in Table 1. It should be stressed that the majority of polymers, given in Table 1, are chemically inert. One can conclude that the adhesion of hydrophobic polymers to primed tissues depends mainly on hydrophobic interactions which exclude chemical bonding of the tissue and polymer. These results illustrate that the adhesion to living tissues is characteristic not only for certain classes of polymers, e.g., cyanoacrylates or polyurethane prepolymers, but for a variety of polymers (see Table 1).

Thus, priming the surface with a certain amount of benzoyl chloride which chemically enables to adjust the degree of adhesion. As stated by the present authors, this

Table 1. Adhesion of synthetic polymers to connective tissue under the effect of benzoyl chloride

Chemical composition of adhesive	Adhesion in units psig[a]	
	Tissue primed with benzoyl chloride	10% benzoyl chloride added to adhesive
30% solution of nitrile rubber in methyl-ethyl-ketone	15.6	13.8
15% solution of poly(vinyl chloride) in ethyl acetate	15.0	
20% solution of polystyrene in ethyl acetate	11.3	10.6
20% silicone elastomer in chloroform (aerosol)	13.1	12.3
Silicone paste (room temperature vulcanizing type)	8.2	
Polyester resin in styrene (peroxide curing)	15.7	
Monomer-polymer mixture of methyl methacrylate (peroxide curing type)	15.0	

[a] The adhesion of methyl-2-cyanoacrylate by this scale is 9.8 psig and the strength of the wound, naturally healed in 10 days, is 10.0 psig.

posibility is very important for preparing adhesives from biologically and chemically inert polymers, because the histotoxicity of this adhesive is determined by the nature and quantity of the monofunctional additive used for modifying the surface. The required amount of benzoyl chloride for priming the tissue surface was determined by the method suggested by the present authors [64]: 15.5×10^{-8} g/cm^2 or 3.3×10^{-11} mole/cm^2, which is harmless for the living tissue. The strength of adhesion was determined one hour after glueing. Indeed, adhesion could occur in this case owing to hydrophobic interactions between the modified tissue and the polymer (e.g., polystyrene). How will the adhesion joint behave several hours or days later? It is very likely that the tissue fluid diffuses along the phase boundary and that the polymer simply delaminates from the surface of the tissue. Below, the case of adhesive delamination in Ringer-Lock solution by the example of glueing catgut with a cyanoacrylate adhesive will be discussed (see Sect. 6).

Elucidation of the mechanism of adhesion to living tissues may lead to the development and application of new adhesives in the future. However, the living tissue is very complex. As stated above, tissue is characterized by mosaic structure composed of alternating hydrophilic and hydrophobic sections. Consequently, hydrophilic and hydrophobic interactions must be considered for glueing tissues. Furthermore, the problem of adhesion has not been solved with respect to engineering. Wang and Bazos [65] clearly formulated some considerations on the mechanism of polymer adhesion to living tissues. These authors state that it is necessary to elaborate a characteristic of adhesion on the basis of general physico-chemical principles in order to approach the solution of this problem. The major problem consists in determining the extent of subordination of bioadhesion to general regularities. According to their opinion, a change in the adhesion of polymers to living tissue, when priming the latter with different amounts of monofuctional compounds of various chemical nature, corroborates the idea that the nature of the polymer and biological tissue have no critical significance for adhesion. In general, this may apply, however, for long-time (days) performance of the adhesive under conditions of high humidity of the living organism this is not the case. Above, the possibility of delamination has been mentioned.

Interaction of the adhesive at the hydrophilic sections of the tissue, when it displaces moisture from these sections, creates a strong adhesion joint, whereupon moisture diffusion at the interface will be less probable than in the case of hydrophobic interactions. Consequently, the edges of the wound will be held under these conditions for a long period of time.

Moreover, it ought to be kept in mind that healing processes should occur in the living tissue after glueing. The structure of the adhesive joint should not prevent these processes, but should even contribute to them under optimum conditions. Consequently, the attainment of strong adhesion is still not a solution of the problem of glueing living (biological) tissues. Let us recall the requirements to medical adhesives listed in the Introduction to this review. Also, it is necessary to stress the fact that the physico-chemical approach to the phenomenon of bioadhesion, elaborated for solving the problem of developing medical adhesives, is very useful and timely. Therefore, it is considered necessary to briefly discuss the mechanism of the formation of an adhesion joint in living tissues.

5 Mechanism of the Formation of an Adhesive Joint in Living Tissues

The existence of various theories of adhesion and the enormous number of publications dealing with the adhesion of polymers to solid surfaces indicates that the problems of adhesion are very complicated.

When considering the adhesion of polymer adhesives to living tissues, the problem becomes still more complicated owing to the complex structure of their surfaces. As stated above, the surface of any living tissue has a mosaic structure, characterized by alternating hydrophilic and hydrophobic microsections with high and low surface energy. The high energy sections are always covered by a fluid coat. The state of these sections is determined by the nature of the tissue and the functions it performs. The latter should always be borne in mind when considering the formation of an adhesive joint. Furthermore, the living tissue features a complex surface relief, evidently, with nonuniformity in size ranging from tens and hundreds of A to tens of microns. The larger-size tens reliefs are significant in forming an adhesion joint. The rheological properties of medical adhesives are also very important; unfortunately they have not been studied sufficiently, to date.

The problems of polymer adhesion and the adsorption theory of adhesion are considered in detail in the monograph by Lipatov [66]. We shall discuss briefly the ideas stated in this monograph that can be used for developing polymer medical adhesives. Information on the structure of living tissue is so scarce that it is impossible to apply the existing ideas of the theory of adhesion. In the literature we have not found any study of phase surface tension of living tissue at the boundary with fluids in view of developing medical adhesives. Phenomena at the phase boundary with a polymer have been investigated until recently only in association with the problem of hemocompatibility, and only the surface energy of the polymers has been determined so far (cf., e.g., Refs. [67,68]). These studies were carried out quite a long time ago.

Physico-chemical investigations have been applied for biopolymers but there is practically no study on adhesion.

The formation of an adhesion joint occurs in several steps [66]: (1) spreading of the adhesive over the surface of the tissue and wetting of the latter; (2) establishment of an equilibrium adhesion contact associated with the molecular properties of the adhesive depending on the properties of adsorption and diffusion; (3) formation of the chemical and physical structure of the adhesive upon curing associated with the appearance of a boundary polymer layer on the tissue surface, differing by its properties from the bulk adhesive (inside the adhesive film). This stage also includes setting (crosslinking) of the adhesive (e.g., in polymerization of cyanoacrylate adhesives), formation of new phases (e.g., gas phase), etc.

The intermolecular interaction between the molecules of the adhesive and those of the tissue surface is possible only in case of molecular contact at the adhesive-substrate interface. Spreading of the adhesive over the surface is determined by the rheological properties of the adhesive, the surface relief, and wetting angle. A thermodynamic equilibrium is attained only rarely under real conditions. Certain theories describe the kinetics of surface wetting determined by capillary forces associated with surface tension. It is possible to calculate the rate of fluid inflow into a narrow or V-shaped slot or a round orifice. The theory considers the rheological properties of the fluids. When the adhesive flows into a slot, orientation of the polymer molecules may occur causing an increase of viscosity. If, in the case of real medical adhesives, polymer molecules form upon spreading, then this will contribute to an increase of viscosity, which is of predominant significance. The intermolecular interactions increase in such a system and relaxation processes are retarded causing the structure of the adhesive joint to be quite far from thermodynamic equilibrium. Hence, it is impossible to predict adhesion within the framework of the thermodynamic theory of adhesion. The occurrence of defects as a result of incomplete wetting is possible owing to the reasons stated above, and there are microsections of living tissue that cannot be wetted by the given adhesive at all. The strict requirements to tissue adhesives are quite understandable. In practice, the temperature can neither be increased to accelerate relaxation processes, nor can a solvent be introduced to reduce intermolecular interactions in the polymer.

The adsorption forces acting at the interface and the factors mentioned above determine the strength at the boundary layer in the adhesive joint. The theory of weak boundary layers has been advanced first by Bikerman [69]. Living tissues are not solid bodies of the type found in most technical glueing (the only exception is bone tissue) and, therefore, the interphase tension will form a "boundary" layer in the tissue. The cells become deformed to a certain depth. This deformation is intensified by chemical bonding of the adhesive by reactive functional groups of molecules at the tissue surface. Moisture should be removed from the interface to ensure contact at the molecular level; this also influences the form of the living cells on the surface. It is necessary to know the permissible values of these changes, such information is presently not available, however.

Next, we would like to stress that adhesion is not linked to the strength of the polymer, determined by various methods (Ref. [66], p. 142). Destruction of an adhesion joint never occurs strictly at the phase boundary. If the microheterogeneity of energy states of a living tissue surface is considered, then an adhesion joint can be presented

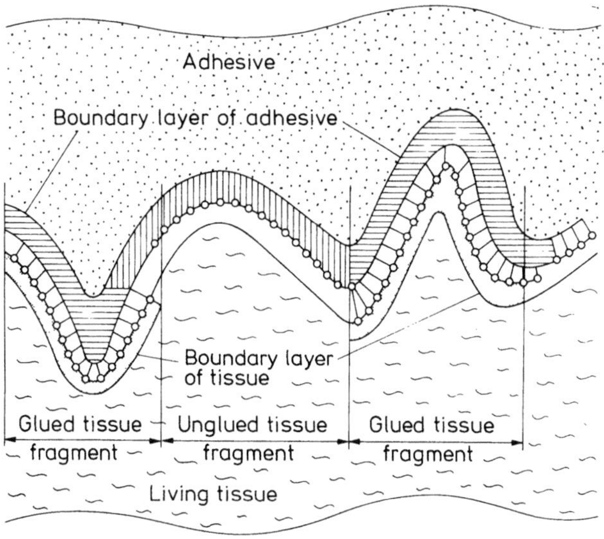

Fig. 1. Section of adhesive joint (schematic)

schematically as follows (Fig. 1). Certain parts of the living tissue surface do not participate in the formation of an adhesion joint. The boundary layers of the adhesive (transverse hatching) and living tissue are shown in Fig. 1.

In summary, the following points are to be stressed which should be accounted for when developing polymer bioadhesives:

1. Consideration of physico-chemical phenomena occurring upon glueing: Investigations of living tissues, estimation of surface energy, surface microstructure at different size levels, the presence of functional groups capable of chemical interaction are the main points characteristic for the physico-chemical approach.

2. The adhesive joint should have a structure which favourably contributes to the healing process.

3. Investigations should be made on the processes occurring at the polymer-living tissue interface from the physico-chemical and biological point of view and ties between physico-chemical and biological phenomena and processes should be determined.

Now, an attempt will be made to demonstrate how we managed to elucidate these relations for the polyurethane glue KL-3.

6 Polyurethane Adhesive KL-3

Polyurethane adhesive KL-3 [70] is a mixture of prepolymer of toluylene diisocyanate (TDI, a mixture of isomers 2,4 and 2,6) with polyoxypropylene glycol with an excess of TDI. The accelerator of curing is dimethyltri(aminomethyl) phenol. The prepolmer containing TDI is mixed before use with the accelerator. The amount of accelerator predetermines the curing time. Thus, the surgeon can adjust the curing time depending on the surgical situation. An aerosol container has been specially developed for adhesive KL-3, wherein mixing of the accelerator and base occur by means of a special device inside the container. It is impossible to change the concentration of the accelerator in this case. However, when using the aerosol form of adhesive KL-3, it is not necessary to do so, because the adhesive is applied on the surface of an open wound. The adhesive cures under conditions of a moist operative area. The chemical structure of a fragment of the polymer network, which forms upon curing, is as follows:

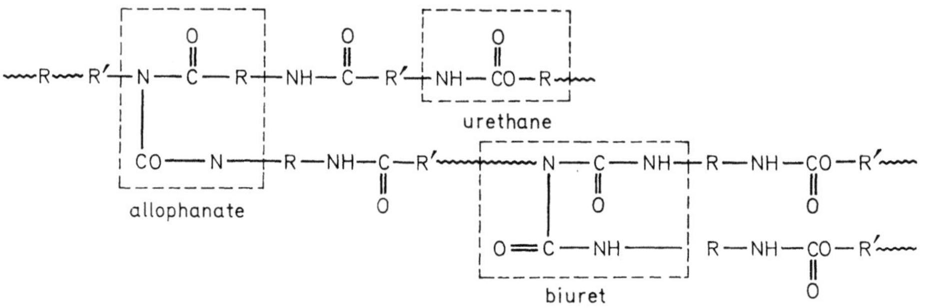

Because of the interaction of adhesive components with water (moisture on wound surface), urea groups are formed and carbon dioxide is released, which causes foaming and the formation of a fine porous structure in the joint with a certain proportion of open and closed pores. Reliable tightness of the adhesive joint is preserved. The glass temperature of the network polymer lies below 0 °C which is characteristic of crosslinked polyurethane elastomers [71]. Consequently, the polymer network is in a viscoelastic state at body temperature, so that certain reversible deformation is possible which is very important for a living organism. The presence of the groups given in the scheme has been confirmed by spectral and chemical methods (Ref. [15], p. 63). The strength of glueing with the adhesive KL-3, determined for aponeurotic sheets [23] by the method described in Section 1, is not inferior to the strength of glueing with the cyanoacrylate adhesive MK-2. It is interesting to compare the strength of glueing of catgut fibers with the adhesives KL-3 and MK-2 when submerging the fibers in Ringer-Lock solution after glueing. The strength of glueing drops by a factor of three after 2–3 hours in the case of MK-2 and remains practically unchanged when using KL-3. Hydrolytic degradation of MK-2 in such a short time is excluded. Consequently, owing to the rigidity of the MK-2 polymer, it adheres poorly to the catgut surface; also, moisture diffused easily into the interphase zone causing delamination. Adhesive joints operate in the organism under conditions of high humidity. In order to attain a strong polymer, investigations of the properties of the adhesive should be started by introducing it into the animal organism and registering the organism's reaction to the adhesive.

6.1 Degradation of Adhesive KL-3 in Model Media and in the Animal Organism

Investigation of a polymer material under real conditions of glueing living tissue is extremely difficult owing to the infiltration of cellular elements, its intergrowth in connective tissues, and other factors. To avoid these complications in case of long-term residence of the sample in the animal organism, the polymer was extracted every three days and implanted in another animal. In this way, the intergrowth of the polymer by connective tissue has been prevented and "pure" samples could be investigated. The polymer was formed under the effect of the living tissue surface, but surface contact with the adhesive was disturbed; nevertheless, the chemical and structural changes in the polymer occurred under the effect of the organism. It is interesting to compare the effect of the organism with that of different media simulating the conditions in the organism.

A wide range of media has been used for investigating adhesive KL-3 (Ref. [15], p. 60): extracts of rabbit liver, kidneys, and skeletal muscles; 0.01% solutions of trypsin, chymotrypsin, and *Streptococcus griseus* proteinase; saline and Ringer-Lock solution. The changes in KL-3 films have been investigated by infrared spectroscopy [72], and the effective cross-linking density using the Flory-Rehner method. When the polymer was placed in saline and Ringer-Lock solution, changes in the infrared spectra were recorded about 15 weeks later. The changes were represented by a slight decrease of the absorption of the band at 1780 cm^{-1}, characteristic of the carbonyl group, the band at 1110 cm^{-1} (valence vibrations C—O of ether bond), and at 1230 cm^{-1} (ester group in urethane —NH—CO—O—). The band at 2960 cm^{-1} was chosen for calculating the relative absorption as an internal standard. The relative optical density was then calculated as a ratio of integral optical densities of the respective bands and bands of valence vibrations C—H (2960 cm^{-1}). Similar changes have been observed in infrared spectra of samples in tissue extracts only after 32 weeks of residence in the media. Proteolytic enzymes (trypsin, chymotrysin, and proteinase) as well as kidney extracts retard considerably the rate of hydrolysis of ester bonds in the urethane groups, while the rate of splitting of ether bonds remains practically unchanged (Figs. 2 and 3). The dependence of changes of the relative absorption at 1230 and 1110 cm^{-1} are presented in Figs. 2 and 3. A comparison of the slopes of the time dependence for each absorption band enables to estimate the influence of the medium on the respective group. Some results of these investigations are presented in Table 2, which shows that proteolytic enzymes have practically no effect on splitting of ether bond. It follows that the polymer obtained from the adhesive KL-3 is subject to simple hydrolytic splitting of ester bonds in model media. A comparison of the results in model media with the results obtained in an organism indicates convincingly that hydrolysis occurs in the organism as well, but the rate of this process is considerably greater (see Figs. 2 and 3, curve 9).

Let us now consider the retarding action of proteolytic enzymes on hydrolysis. Attention should be drawn to the fact that the carbonyl absorption band splits into two parts as a result of interaction of the cured KL-3 with proteolytic enzymes and kidney extract. Evidently it is associated with the specific interaction of the enzyme and urethane group in the polymer, the structure of which resembles the peptide group of a protein molecule. Owing to the specific action of the enzyme, this interaction does not accelerate the hydrolysis of urethane groups but even retards it owing to the shielding effect of the enzyme protein molecule.

It follows from the structure of cured KL-3 that secondary reactions are possible in the formation of the network manifested in the formation of allophanate and biuret groups. The presence of these groups and the change in their occurrence under the influence of model media and of the animal organism has been studied by their aminolysis with diethylamine up to the point where they have become completely transformed into urea groups, which were determined by IR spectrometry at 1655 cm^{-1} (Ref. [15], p. 63). It has been found that allophanate and biuret bonds disappear in the polymer kept 15 weeks in Ringer-Lock solution and after 12 days of implantation in the organism.

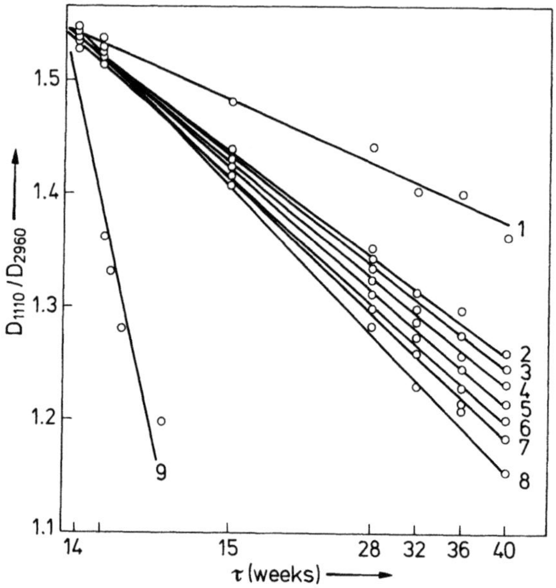

Fig. 2. Change of relative band absorption 1110 cm^{-1} with time under effect of organism (9) and model media: 1 0.01% proteinase solution, 2 chymotrypsin solution, 3 0.01% trypsin solution, 4 extract of rabbit kidneys, 5 extract of rabbit muscles, 6 extract of rabbit liver, 7 Ringer-Lock solution, 8 saline

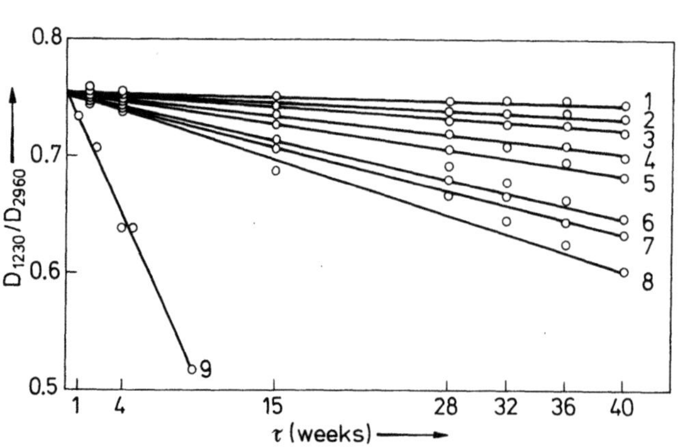

Fig. 3. Change of the relative integral band absorption 1230 cm^{-1} with time under the effect of the organism and model media (for designation of curves see Fig. 2)

Table 2. Conditional time change of relative integral intensity (tgα) of absorption bands in IR spectra of adhesive KL-3 subjected to model media and the animal organism

Medium	tgα	
	1110 cm^{-1}	1230 cm^{-1}
Saline	0.133	0.051
Ringer-Lock solution	0.135	0.041
Extract of rabbit liver	0.103	0.037
Extract of rabbit muscles	0.100	0.025
Extract of rabbit kidneys	0.120	0.017
0.01 % trypsin solution	0.110	0.009
0.01 % chymotrypsin solution	0.116	0.006
Animal organism	0.594	0.385

The effective crosslinking density of the polymer KL-3 (Fig. 4) decreases with increasing interaction time in model media and most intensively under the influence of the animal organism. The change in the effective crosslinking density is in good agreement with the results of spectral investigations of polymer KL-3 degradation in model media and in the animal organism.

Thus, the mechanism of polyurethane adhesive degradation under the influence of model media and in the animal organism is similar and consists in hydrolytic splitting of ester bonds in the urethane groups and in slower hydrolysis of ether bonds. The degradation process occurs much faster under the effect of the organism. Again, it should be recalled that the cellular mechanism of biodegradation is excluded through the transplanting of cured samples to new animals every three days. The latter mechanism will be discussed below.

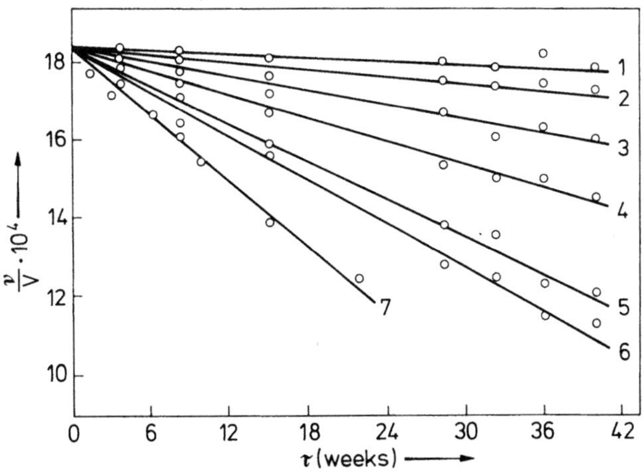

Fig. 4. Dependence of the effective crosslinking density of cured adhesive KL-3 on time for different media: 1 proteinase solution, 2 extract of rabbit kidneys, 3 extract of rabbit liver, 4 extract of rabbit muscles, 5 Ringer-Lock solution, 6 saline, 7 animal organism

It became necessary to investigate the morphology of the cured adhesive and its changes. The changes occur owing to splitting of chemical bonds in the polymer network.

6.2 Morphology of the Cured Adhesive KL-3 and its Change During Biodegradation

Investigation of the supermolecular organization in an adhesive and its changes in the process of biodegradation is very important for comprehending the processes occurring in the polymer during biodegradation. The supermolecular like organization of the cured KL-3 has been investigated by electron microscopy [73].

Cured KL-3 has been shown to be characterized by — globular formations [73].

During degradation, the character of supermolecular formations remains unchanged and globular-like while the size of the globules changes. The size distribution of the globules has been calculated with data from electron microscopy (Fig. 5). The initial sample (curve 1) is characterized by a maximum at 800–1000 Å. Only a small fraction of globules was found above 4000 Å. A certain divergence of the curves and shift of the maximum to larger globules appears to be a result of the effect of the model media (curves 2–4). The influence of saline on the alteration of the supermolecular organization (curve 2) is stronger compared with other model media. The chymotrypsin solution renders the least effect on the polymer (curve 5). These results are in good agreement with data of chemical measurements and of the polymer crosslinking density.

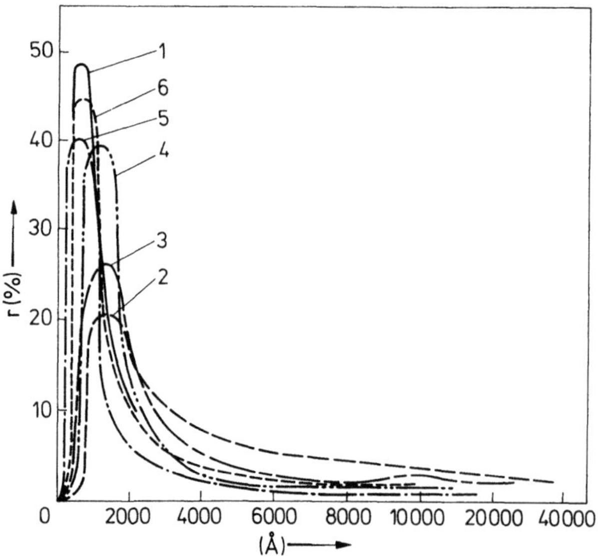

Fig. 5. Size distribution of globular formations in the adhesive KL-3 cured *in vitro* (1), after 90 days in model media: saline (2), extract of rabbit media (3), extract of rabbit liver (4), animal organism (26 days) (5), chymotrypsin solution (6)

Attention should be drawn to the change in the supermolecular organization of a polymer subjected to the effect of saline. A considerably large number of large globules (up to 13000 Å) appears in the structure (curve 2). Fewer globules of this size are observed in the structure of a polymer subjected to the effect of muscle extract (curve 3). The changes in the supermolecular organization of a polymer subjected to different media are in correspondence with changes in structure and crosslinking density. Saline has been found to be the most active hydrolyzing medium. The maximum rate of polymer degradation has been observed within the living organism. The specific influence on the supermolecular organization of the polymer consists in the disappearance of globules being larger than 3000–4000 Å, i.e. formation of large globules does not occur. This result requires some explanation. The observed effect cannot be associated with the influence of cellular elements as the experiment excluded the possibility of their penetration into the polymer. In order to understand the activity of the living organisms it was necessary to investigate the effect of mechanical loads on the degradation of the polymer in saline solution, i.e., in a medium which in the formation of the largest globules was recorded.

6.3 Influence of Static and Dynamic Loading on Degradation of the Cured Adhesive

The cured KL-3 adhesive was subjected to static and dynamic loading in saline at 37 ± 1 °C. A stress was applied at constant tension, which was approximately equal to 20% of the breaking stress, and the sample was held in a stressed state up to 90 days. In certain time intervals, the samples were compared with those submerged freely in the medium. The polymer was subjected to dynamic loading at 18 kHz. The sample being 20 × 10 × 5 mm in size was supported as a cantilever with a 0.3 mm vibration amplitude. The results indicated that superposition of force fields causes qualitatively similar changes in the IR spectra compared to spectra of samples submerged freely in saline. A decrease of optical densities was monitored for the same absorption bands at 1230, 1730, and 1110 cm^{-1}. However, their time dependence was different. The decrease of the absorption intensity at static loading is faster by a factor of 5–7. The change of the relative optical densities as a function of the number of loading cycles is represented by a straight line.

The type of supermolecular organization of cured adhesives has not changed after long-term loading and remained globular-like. A considerable shift of the maximum on the size distribution to the region of smaller globules occurs under the influence of static and dynamic loads (Fig. 6). Large globules with a diameter of 1500–2000 Å practically disappear. Unlike the polymer submerged freely in solution (Fig. 5), large globules are not formed under the action of force fields on the polymer. This difference in the change of the sizes of the supermolecular formations in the same medium may be associated with the different rate of hydrolysis and diffusion of the degradation fragments in the medium and with the different conditions of forming supermolecular formations. Degradation of the polymer network is indicated by a decrease of the value of the stress at break of the samples depending on the static loading time and the crosslinking density. The effective crosslinking density decreased approximately by 20% after a 90-day loading and the breaking strength fell by 10%. These results confirm the presence of cleavage of chemical bonds in the crosslinked polymer.

These results allow us to understand the physico-chemical effects of a living organism on an adhesion joint and on a polymer in a living organism.

In addition, the cellular mechanisms of biodegradation and adhesive resorption are very important and, therefore, we shall discuss this point in more detail.

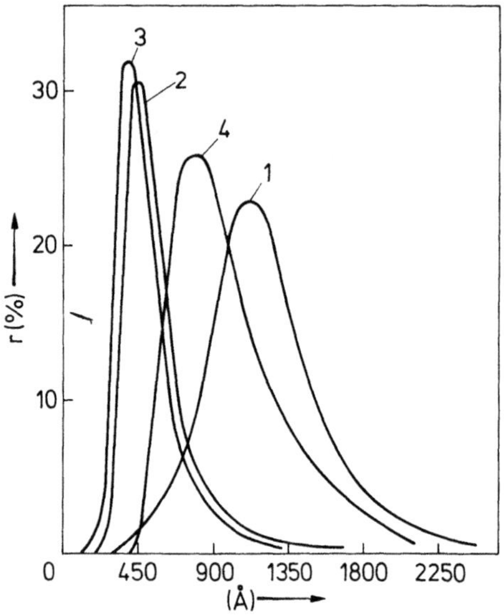

Fig. 6. Size distribution of globular formations in the cured adhesive KL-3 (1) and after static loading for 50 days (2), 90 days (3); after dynamic loading for 4 min. at 18 KHz (4)

6.4 Cellular Mechanism of Biodegradation and Resorption

The possibility of polymer biodegradation in the organism by hydrolysis has been illustrated by the example of polyurethane adhesive KL-3 [15]. It is evident that all the polymers that can undergo hydrolysis (polyethers, polyamides, polycarbonates, etc.) will also be destroyed in the organism by hydrolysis. This was confirmed by studies of polycaproamide in animals [75]. Besides hydrolysis, there exists a cellular mechanism of polymer degradation in the organism. Histological degradation of cyanoacrylates has been observed in [76] where it was possible to see tiniest particles owing to intensive and bright dyeing of polymer. However, the cellular mechanism of polymer degradation in an organism has not been studied in detail. It would be very interesting to determine what cells participate and how they do so in the degradation and resorption of the adhesive layers.

As a result of checking a series of dyes, it has been found that the dye Sudan III–IV (mixture) is sorbed delectively by KL-3 preparations. The intensity of dyeing to a reddish-orange color is so strong that it is possible to determine particles of the adhesive in the preparation up to a size of 3 μm (Ref. [15], p. 69). Microscopic investigations of the experimental material have been performed in glueing wound surfaces of different animal organs by the KL-3 adhesive. The results indicated that the tissue changes in the area of contact with the adhesive have been of principally similar character, irrespective of the place of implantation (kidney, liver, small and large intestine, gluteal muscles). The reaction of the organism tissues is described in detail in Ref. [77]. Only the cellular mechanism of KL-3 degradation will be discussed.

Investigations of adhesive resorption have indicated that the process starts soon after implantation and is closely associated with the productive reaction, starting at

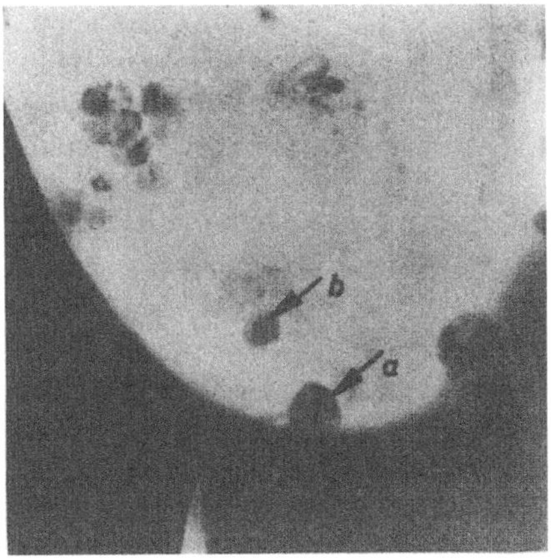

Fig. 7a and b. Fragments of adhesive KL-3 in cytoplasm of macrophages on polymer surface **(a)** and at some distance from it **(b)** 14 days after application on rabbit liver. Selective dyeing with Sudan III to IV, additional dyeing with hematoxylene

the place of adhesive contact with the tissue. Intergrowth of the connective tissue gemmas during the initial stages of the reactive process (14 days after the operation) spreads from the connective capsule into the cells (pores) which are in contact with the adhesive, owing to degratation of the pore walls. The fact that gigantic cells and macrophages active in phagocytosis have not been detected in these areas suggests that filling of the pores by connective tissue has been preceded by degradation of their walls by hydrolysis and infiltration of tissue liquid containing leukocytes and macrophages. The function of the macrophages was to destroy the surface of the adhesive pore they were in contact with. The macrophages in the microphotograph (Fig. 7) contain adhesive granules that are on the surface of the adhesive and those that have been removed. The number of macrophages increased every 2–3 months after the start of the experiments, gigantic cells of foreign bodies appeared in the incorporating connective tissue. Polymer particles have not been found there. The cells were located in depressions that formed on the polymer surface in contact with the tissue. Many gigantic cells penetrated the polymer cavity six months later. The process may be understood as follows.

At first, a small depression (a lacuna) forms in the part of the polymer neighbowing the gigantic cell to where a minor amount of cytoplasm and one, rarely two, nuclei are displaced simultaneously. The size of the lacuna increases owing to the lysis by the gigantic cell and a greater amount of cytoplasm displaces there to, while the number of nuclei increases by a factor of 5–6. When the cavity grows large enough, the gigantic cell enters it completely, its outgrowths penetrate deeply into the newly formed cavities in the polymer mass (Fig. 8). The number

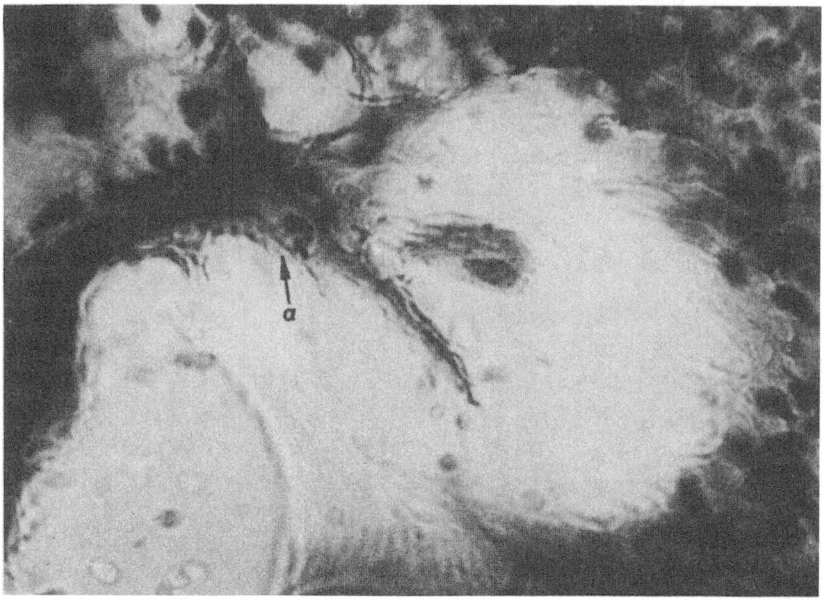

Fig. 8. Penetration of numerous cytoplasmic outgrowths of gigantic cells of foreign bodies into the polymer KL-3 (a); 3 months after glueing intestinal loops in rabbit (azure-II-eosine by Maximov)

of gigantic cells increases along with the amount of newly formed connective tissue, dividing the polymer mass into fragments. Consequently, the gigantic cells in penetrating into the polymer and increasing their size, destroy the polymer owing to their lytic activity. The results indicate that degradation of polyurethane adhesive KL-3 occurs as a result of a combination of various processes. The primary means of degradation, occurring continuously while the polymer is in the organism, is hydrolysis under the influence of tissue liquids and interstitial substances in the connective tissue. Disintegration of the polymer occurs at a later time owing to the phagocytizing activity of the macrophages and lysis by gigantic cells of foreign bodies.

The elimination of the products of biodegradation of cyanoacrylate adhesive has been discussed in Section 3. The adhesive KL-3 also has been investigated to determine process elimination using ^{14}C labeling by introducing ^{14}C-ethyl alcohol into the curing adhesive. The amount of the introduced alcohol was calculated by assuming that it will react with 7% of the isocyanate groups of the adhesive [78]. The results indicated no cumulation of radioactive fragments in the organs and tissues (Ref. [15], p. 73). Following the elimination through the respective outlet tracts and determining the specific radioactivity of tissues of different organs, the authors compared these data with the results of histological and chemical data.

Elimination of a considerable amount of radioactive fragments has been observed in the first 5–7 days, then the specific radioactivity of the blood serum, urine, and feces decreased greatly by the tenth day (Fig. 9). The weight of the polymer sample decreased approximately by 5% in 5 days. Determination of the soluble fraction (6–7%) of the initial polymer sample upon extraction with toluene made it possible to conclude that the soluble fraction of KL-3 was eliminated from the organism first during the initial five days when the connective capsule had not yet been formed. The decrease of the specific radioactivity (Fig. 9) by the tenth day after implantation is associated with the formation of the connective capsule by that time, owing to which the rate of adhesive degradation is retarded, dropping gradually almost to zero after six months.

The results of determining the radioactivity of the brain, skeletal muscle, heart, spleen, liver, lungs, kidneys of rats after introducing the polymer adhesive KL-3 are in good agreement with the data presented in Fig. 9. The specific radioactivity of the investigated tissues decreases sharply by the tenth day, and the process of adhesive joint degradation is retarded. A certain trend toward an increase of radioactivity in some tissues is observed after two months of implant residence in the animal organism. This period is characterized under the conditions of the histological experiment by intensive fragmentation of the adhesive by connective elements.

Threee stages may be observed in the biodegradation of the polymer adhesive KL-3. The first stage in the course of the first week is characterized by a high rate of elimination of adhesive fragments from the animal organism. Their amount is affected by the introduction of a monofunctional additive (ethyl alcohol) during the preparation of the adhesive. The large contact area of the adhesive with the wound surface contributes greatly to this process. The second stage (up to two months) is characterized by retarded elimination of the degradation products and represents the process of polymer biodegradation. Healing-regenerative organization of the wound and incapsulation of the adhesive occur simultaneously. The duration of this period may vary considerably depending on the place of introducing

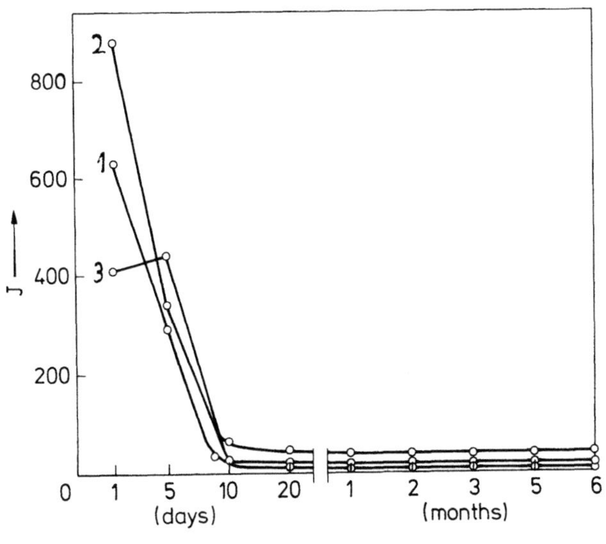

Fig. 9. Specific activity of blood serum (1), urine (2), feces (3) after introducing labeled adhesive KL-3 into organism

the adhesive into the organism, as well as on its amount and on the animal species. Insignificant acceleration in the elimination of the products of biodegradation is typical of the third stage and is manifested by the level of radioactivity in certain organs; however it is not related to the level of radioactivity in the blood, feces, and urine (Fig. 9). This period is associated with fragmentation of the adhesive by connective elements. It is characteristic for this experiment but not obligatory for all other cases.

Evidently, all these stages will be displayed differently with adhesives of various chemical nature and structure on the microscopic scale.

6.5 Reaction of Organism Tissues with KL-3

We shall describe very briefly the reaction of tissues to the adhesive. It is necessary to note that the method of adhesive application on an uninjured organ has been used for determining the degree of local histotoxic influence of the adhesive KL-3. This method has been used for investigating the liver, kidneys, small and large intestine, and gluteal muscles in rabbits and white rats. Summarizing briefly the investigations, it can be said that KL-3 forms a fine porous film of high elasticity in the organism. The cured adhesive is in a close contact with the tissue causing aseptic inflammation with a characteristic change of the mycotic, macrophagal, and fibroblastic stages. The releasing products of polymer degradation cause no appreciable toxic effect in the surrounding tissues. An increased content of nucleic acids has been found in the tissue elements around adhesive fragments. This indicates the absence of toxic effects by the products of biodegradation. The products of biodegradation are subjected evidently to rapid inactivation. Minor changes in the tissues to which the KL-3 adhesive has been applied, and regeneration of their state one month after the operation lead to the conclusion that the histotoxicity of KL-3 is minimal.

The biochemical indices of skeletal muscles in plastic surgery of injuries have been investigated in detail for KL-3, because muscles are injured quite frequently (Ref. [15]; p. 81).

Investigation of the antigenic composition of skeletal muscles at closing wounds by means of the adhesive (Ref. [15], p. 89) has indicated that the adhesive causes no change in their antigenic behavior. These results are quite significant in case of a repeated use of the adhesive in the same organism. Also, it is interesting to note that KL-3 is autosterile and ensures intensive hemostasis, though the reasons of this are not yet clear.

Using the example of the adhesive KL-3, we have raised a series of questions that ought to be answered when developing an adhesive for glueing living tissues. The conclusions can be extended for other polymeric materials capable of hydrolytic degradation. Owing to the great microporosity of the adhesive joint, KL-3 features an expanded surface contact with the media of the organism, hence, the physico-chemical phenomena at the interface are of great significance. The porous structure of the polymer is very important also for the responsive reaction of the organism. Therefore it is necessary to discuss the phenomena at the biocompatible polymer-living tissue interface.

7 Phenomena at the Biocompatible Polymer-Living Tissue Interface

For glueing living tissues, the hydrophilic sections of the tissue should be selected because these sections have high surface energy and adhere well to the cured adhesive. It is necessary for the adhesive to pass through the stage of polymer formation and displace moisture from the tissue surface possibly reacting with water. Reactive surfactants are formed in this case. The principles of using reactive surfactants in adhesives has been discusssed by us with respect to glueing [50]. The reactive surfactant improves wetting in the first stage of adhesive interaction. The reactive surfactant gradually reacts to form a network and is substituted on the surface for the cured adhesive. The adhesive KL-3 also passes through a stage which is associated with a rise of surface activity owing to the formation of urea groups that contribute to its spreading over the surface. Evidently, there is a difference between glueing of living tissues and technical glueing, in which chemical bonding may cause undesirable effects [79]. When glueing living tissues, liquid diffusion may occur easily at the adhesive-tissue phase boundary (see Sect. 9) as well as delamination of the adhesive from the surface. The adhesive KL-3 with its NCO groups can react with mobile hydrogen atoms in protein molecules. The possibility of isocyanate groups reacting with NH_2 groups of protein molecules has been illustrated by the example of KL-3 interaction with chymotrypsin [80]. Furthermore, the deformation of the soft tissue structure is also important.

The fine porous structure of the adhesive joint causes an early degradation by the cellular mechanism. For KL-3, under experimental conditions, the cellular mechanism of degradation manifests itself two weeks later. Intensive degradation of the adhesive by the hydrolytic mechanism is due to the highly expanded surface under formation of cavities and subsequently of gigantic cells of foreign bodies.

Degradation of bulk biocompatible polyurethanes occurs mainly by the hydrolytic mechanism and the stage of cellular degradation is initiated only at later stages when cavities appear as a result of polymer surface erosion (Ref. [15], p. 44).

All conclusions made above are equally valid for all hydrolyzing, fine-porous and bulk biocompatible polymers.

8 Conclusions

In summary, it should be stated that attachment of adhesives to living tissues have not been studied in detail. The strength of the adhesive joint should be determined experimentally on living tissues under conditions that are close to a real surgical situation, as existing models are insufficient. Investigations of the phenomena at the adhesive-living tissue interface from the physico-chemical as well as the biological points of view are still needed.

It is impossible to develop a versatile adhesive for all kinds of living tissues, as these differ in their physico-chemical characteristics, functions, rate of regenerative processes, type and degree of saturation with enzymes, etc. It will be necessary in the near future to synthesize and investigate several types of medical adhesives.

9 References

1. Lipatov, Yu. S.: Physical chemistry of filled polymers. Moscow, "Khimiya" 1977
2. Lipatova, T. E., Pkhakadze, G. A.: Polymers in endoprosthetics. Kiev, Naukova Dumka 1983
3. Oppenheimer, B. S., et al.: Europ. Cancer Res., *15*, 333 (1955)
4. Harrison, J. H.: Amer. J. Surg. *95*, 3 (1958); *95*, 16 (1958)
5. Payne, J. T.: Amer. Surg. *36*, 615 (1970)
6. Nelson, R. A. et al.: Fed. Proc. *27*, 707 (1968)
7. Coover, H. W. et al.: Soc. Plastics Eng. J. *15*, 413 (1959)
8. Coover, H. W. et al.: Cohesive News *1*, 1 (1961)
9. Ota, K. et al.: Arch. Surg. *96*, 231 (1968).
10. Matsumoto, T. et al.: Milit. Med. *134*, 247 (1969)
11. Niololerc, M. et al.: Meditter. Med. *2*, 543 (1974).
12. Herrman, G. B., Woodward, S. C.: Surgery No 4, 559 (1966)
13. Kósa, L., Banyi, E., Lang, T.: Plastubol EGYT, Nr. 1716/A, Pharmaceutical Works, Budapest
14. Korshak, V. V., Polyakova, A. M., Shapiro, M. S.: J. D. 1. Mendeleev All-Union Chem. Soc. *24*, No. 1, 51 (1969).
15. Lipatova, T. E., Pkhakadze, G. A.: Application of polymers in surgery. Kiev, Naukova Dumka 1977
16. Gigauri, V. S. et al.: Khirurgiya No. 10, 101 (1979)
17. Galperin, E. I. et al.: Khirurgiya No. 7, 41 (1978)
18. Leonard, F. et al.: J. Appl. Polym. Sci. *10*, 259 (1966)
19. Heis, W. H., Guthy, E., Becker, H. M.: Lang. Arch. Klin. Chir., *208*, 793 (1964)
20. Woodward, S. C. et al.: Ann. Surgery *162*, 113 (1965)
21. Braunwald, N. S., Awe, W. C.: Surgery *51*, 817 (1962)
22. Thiel, K. et al.: Lang. Arch. Klin. Chir. *314*, 62 (1966)
23. Burenko, G. V., Pkhakadze, G. A.: Klinicheskaya Khirurgiya No. 4, 60 (1971)
24. Volkov, M. V., Perepelkin, V. P., Yanovskaya, E. M.: Materials of 2nd Rep. conf. on Appl. of Polymers in Surgery. Kiev. 1969, p. 96
25. Volkov, M. V. et al.: Polymers in medicine (Poland), *2*, 63 (1972)
26. Yanovskaya, E. M.: Eksp. Khirurgiya Anesteziologiya, No. 6, 31 (1969)
27. Reynolds, R. C. et al.: J. Surg. Res., *6*, 132 (1966)
28. Lewers, D. T., Just-Viera, G. O., Geager, G. H.: Surgery, *87*, 627 (1963)
29. Collins, C. G. et al.: Arch. Surg., *93*, 428 (1966)
30. Pani, K. C. et al.: Surgery, *63*, 481 (1968)
31. Herrman, J. B., Woodward, S. C.: Surgery, *59*, 559 (1966)
32. Timokhina, V. I. et al.: All-Union Symp. "Synthetic polymers of medical use", Tashkent, Acad. Sci. Uzbek SSR, 1973, p. 56
33. Timokhina, V. I. et al., In: Polymers in medicine. Kiev, Naukova Dumka, 1976, p. 43
34. Yegorov, V. I., Boguslavskaya, E. S.: Eksp. Khirurgiya Anesteziologiya, No. 2, 46 (1971)
35. Yashon, D. et al.: J. Neurosurg. *25*, 883 (1966)
36. Matsumoto, T. et al.: Milit. Med. *134*, 247 (1969)
37. Lipatova, T. E., Pkhakadze, G. A.: Medical adhesives. Kiev, Naukova Dumka 1979
38. Koenlein, H. E., Lempele, G.: Surgery, *66*, 377 (1969)
39. Tatooles, C. J., Braunwald, N. S.: Surgery, *60*, 857 (1966)
40. Braunwald, N. S., Gay, W., Tatooles, C. J.: Surgery *59*, 1024 (1966)
41. Matsumoto, T., Pani, K. C., Heisterkamp, C. A., Hamit, H. T.: Milit. Med. *134*, 98 (1969)
42. Wenning, J., Geisller, G.: Polymery w medycynie (Poland), *3*, 73 (1973)
43. Mohammed, A. I.: Klinicheskaya Khirurgiya, No. 3, 23 (1976)
44. Lidarik, M., Kincl, I.. Roth, V., Bring, A.: SNTL, 2nd ed. Prague, 1961, p. 289
45. Golovin, G. V.: Mestnaya Khirurgiya, *83*, No. 7, 45 (1959)
46. Volkov, M. V.: Khirurgiya No. 2, 3 (1966)
47. Lipatova, T. E., Veselovsky, R. A., Pkhakadze, G. A.: Adhesive for gluing soft body tissues. Can. Pat. 1096097, (1981), US Pat. 4,057,0535, (1977), BRD Pat. 2617404, 1982 Fr. Pat. 23508737, (1978)
48. Wang, P. Y., In: Biomedical Application of Polymers, Polymer Science and Technology, New York: Plenum Press, 1975, vol. 7, p. 111

49. Veselovsky, R. A., Lipatova, T. E., Lipatov, Yu. S.: Dokl. Akad. Nauk SSSR 233, 1146 (1977)
50. Veselovsky, R. A., Vysotskaya, G. V., Lipatov, Yu. S.: Dokl. Akad. Nauk SSSR 248, 915 (1979)
51. Salvatore, G. F., Mandarino, M. P.: Ann. Surg. *149*, 107 (1959).
52. Salvatore, G. F., Gimer, W. S., Kashgarian, M.: Surg. Gyn. Obst., *112*, 463 (1961)
53. Drompf, B. W.: Amer. J. Surg., *99*, 733 (1960)
54. Wondrak, E.: Čas. Lek, Čes. C1, 1155 (1961)
55. Redler, I.: J. Bone Joint Surg. 44, 1621 (1962)
56. Bernhard, W. F. et al.: Circularion *27*, 739 (1963)
57. Cooper, C. W., Grode, G. A., Falb, R. D.: In Tissue Adhesive in Surgery, ed. Matsumoto, T., p. 192, Flushing: Med. Exam. Publ. Co., 1972
58. Grode, G. A., Cooper, C. W., Falb, R. D.: New Scientist *36*, 27 (1967)
59. Llevellyne-Thomas, F., Wang, P. Y., Cannon, G. S.: J. Biomed. Mat. Res. *8*, 35 (1974)
60. Llevelyne-Thomas, F., Wang, P. Y., Vinals, J.: Biomat. Med. Dev., Art. Org. *1*, 507 (1973)
61. Matsumoto, T., et al.: Jinkō Zōki, 5, 152 (1976)
62. Galligun, J. D., Minor, F. W., Schwartz, A. N.: Adhesion in Biological Systems, p. 258, ed. R. S. Manalg, New York: Academic Press 1970
63. Wang, P. Y., Llevelyne-Thomas, F.: J. Biomed. Mat. Research, *12*, 831 (1978)
64. Wang, P. Y., Forrester, D. H.: Trans. Amer. Soc. Artif. Int. Organs *20*, 504 (1974)
65. Wang, P. Y., Bazos, M. J.: In Physicochemical Aspects of Polymer Surfaces, New York: Plenum Publ. Corp. 1983 Vol. 2, p. 943
66. Lipatov, Yu. S.: Colloidal Chemistry of Polymers. Kiev, Naukova Dumka 1984
67. Bruck, S. D.: J. Polym. Sci.-Polym. Symp. N 66, p. 283, 1979
68. Andrade, J. D. et al.: J. Polym. Sci., Polym. Symp. *00*, 313 (1979)
69. Bikerman, J. J.: The Science of Adhesive Joints. N. Y.—London, Acad. Press, 1968
70. Lipatova, T. E.: J. Polym. Sci., Polym. Symp., No 66, 239 (1979)
71. Lipatov, Yu. S. et al.: Rubber Chem. Technol. *41*, 537 (1968)
72. Lipatova, T. E. et al.: Polymers in medicine (Poland), *4*, 313 (1974)
73. Lipatova, T. E., Novikova, O. E., Pkhakadze, G. A.: Dokl. Akad., Nauk Ukrainian SSR, B, No. 9, 819 (1975)
74. Lipatova, T. E., Veselovsky, R. A.: Vysokomol. soedin. A*11*, 1459 (1969)
75. Voronkova, O. S. et al.: Vysokomol. soed. A*13*, 677 (1971)
76. Miko, I.: Polymery w medycynie *9*, 39 (1979)
77. Kovalev, M. M., Tereshchenko, T. L., Suslov, E. I.: Polymers in medicine (Poland), *4*, 277 (1974)
78. Pkhakadze, G. A. et al.: Trans. 29th All-Union Congress of Surgeons, Kiev, 1975, p. 222
79. Lipatov, Yu. S.: J. Adhesion 10, 157 (1979)
80. Lipatova, T. E. et al.: Ukr. Biokhim. Zh. *51*, 319 (1979)

Editor: K. Dušek
Received November 12, 1985

Heparin-Containing Polymeric Materials

Nicolai A. Platé and Lev I. Valuev

Department of Chemistry, Moscow State University,
Moscow 119899, USSR

Methods for preparing heparin-containing polymeric materials by means of ionic and covalent immobilization of heparin on various polymers are surveyed. The data on the biological activity of heparin are discussed as well as the probable mechanisms of thromboresistance enhancement endowed to polymeric materials by this anticoagulant. Some approaches toward an increased efficiency of anticoagulant properties of immobilized heparin are analyzed, and the position of heparin-containing polymers among other biomedical polymers is discussed.

Advances in Polymer Science 79

1 Introduction

In the last 10–15 years synthetic polymeric materials have become widely used in practical medicine. The important particular field of their application is the production of various endoprosthetic and other devices contacting with blood. Thereby polymer chemists have met the necessity to design and create specific procedures for enhancing thromboresistance of synthetic polymers. One of the most common approaches is provided by the chemical modification of polymers with various compounds inhibiting the blood clotting process. These are pyrolytic carbon [1], a number of hydrophylic monomers [2,3], some surface-active compounds [4], and a wide range of biologically active substances that are potent in effecting thrombosis and/or fibrinolysis processes are the most prominent representatives of these compounds. Initial stages of thrombus formation — i.e., protein adsorption, activation of the contact factors, platelet adhesion, aggregation and activation — are predominantly affected when the modifying compounds are selected from the first three of the groups given above. However, the scope of activity of biologically active compounds is much broader, hence their use for polymer modification provides the possibility to inhibit thrombus formation actually at any stage depending on the nature of the compounds in use. Reported biologically active compounds for this purpose are proteolytic enzymes, e.g., urokinase [5], fibrinolysin (plasmin) and enzymes of bacterial origin [6–9], agents preventing platelet aggregation [10], and a natural blood anticoagulant — heparin.

Of these the last one has been most widely used, since heparin-modified polymeric materials exhibit the highest and by today unsurpassed effects of thromboresistance enhancement. Many of these materials have not only proved to be potent in trials on animals, but have already found clinical application. These achievements have stimulated continuous interest in heparin-containing polymers (HCP) which is best manifested by listing the investigations performed in the field in recent years and still under way. They involve the new procedures for the synthesis of HCP providing minimal loss of activity of bound heparin, the studies of interactions of HCP with blood and its individual components, as well as on the mechanism of enhanced thromboresistance of HCP, and the search for new tasks for HCP.

The year of 1961, when Vincent Gott [11] observed the inhibition of thrombus formation by immobilized heparin for the first time, was marked as the second birth date of heparin, since it was for the first time isolated from liver tissue. Its anticoagulant action was detected in 1892. Although more than 20 years have passed since Gott's publication, there is still much confusion concerning the views on the mechanism of enhanced thromboresistance of heparin-modified polymers, which greatly hinders the introduction of HCP into clinical practice.

The aim of the present review is to systematize and analyze the results of investigations concerning the synthesis and properties of HCP on the basis of literature data and own results. The probable mechanism of the effect of polymer-bound heparin on the blood clotting system will be discussed. A survey of the most promising pathways for further development of the chemistry of HCP involving the development of methods of their synthesis as well as the increase of the activity of immobilized heparin itself will be given.

2 A Brief Survey of the Structure and Properties of Heparin

Heparin is a heteropolysaccharide whose chains consist of alternating derivatives of 2-amino-2-desoxy-*D*-glucose and uronic acid linked together via 1 → 4 glycoside bonds [12].

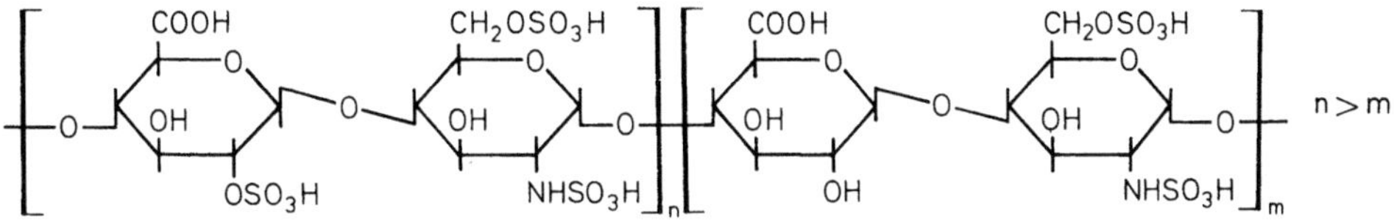

The molecular mass of heparin varies between 10000 and 30000, depending on the source of its isolation [13]. Since the above formula of heparin gives only the approximate presentation of the structure of its macromolecule, let us examine the data concerning functional groups and other components of the heparin macromolecule.

Each disaccharide unit contains 2 or 3 sulfate groups which are attached in the 6th position to the amino groups of *D*-aminoglucose and in the 2nd position to the oxy groups of iduronic acid [14]. N-acetyl groups have also been detected in heparin, the content of N-acetylaminoglucose varying for different species from 2 to 31% [15]. About 10% of the amino groups are free, i.e., are not substituted [16]. Heparin is not strictly a pure heteropolysaccharide: its carbohydrate chain is bound to the oxy group of the serine residue of the proteinaceous component via a neutral polysaccharide chain [14]. The relative content of the carbohydrate and proteinaceous components depends on the source and method of isolation of heparin. Heparin preparations have been reported which do not contain oligopeptides [17]. Some investigators assume that there are several carbohydrate chains attached to a protein molecule [18] in heparin, although a single carbohydrate chain bound to a polypeptide also cannot be totally excluded [14].

A three-dimensional model of heparin was proposed based on the X-ray analysis of "heparin-ionic copper" complexes [19]. According to the model, the carbohydrate chain is a sparce but a regular helix containing four disaccharide units per helix coil. The length of such helical segments is 21 Å, the diameter of the helix being 10 Å. Two thirds of the ionizable groups, including all N-sulfate groups and carboxy groups are located on the inside of the helix. One third of the groups, including the O-sulfate ones, is located on the outside of the helix, thus forming the ionic atmosphere of the molecule.

The activity of heparin is to a large extent determined by the amount and the mode of distribution of the functional groups along the carbohydrate chain. For instance, heparin with a high content of amino groups exhibits lower anticoagulant activity. According to the data reported in Ref. [16], heparin is entirely inactivated when more than a half of the N-sulfate groups are desulfated to free amino groups. Eliminating 50% of the O-sulfate groups [20] as well as the blocking of up to 50% of the carboxy or 10–20% of the hydroxy groups does not result in a reasonable change of heparin activity [21].

The anticoagulant activity of heparin is endowed by its ability to form strong complexes with a variety of blood clotting factors and thus neutralize their action. For instance, heparin interacts with thrombin, fibrinogen, prothrombin, factors IX–XII etc. [22,23]. The anticoagulant activity of heparin is exhibited to a largest extent by the presence of antithrombin III which is a blood glycoprotein with a molecular mass of 65000 and having a carbohydrate content of 9–13.4% [24]. Antithrombin III neutralizes the activated factors IXa–XIIa and other proteases [22]. However, its main function is the irreversible inactivation of thrombin, about 70% of the antithrombin potentialities of the blood plasma being provided by antithrombin III [25]. In the presence of heparin, the inhibitory activity of antithrombin III towards the named compounds rises 50 to 60 fold.

According to the prevailing opinion, the mechanism of the accelerating effect of heparin on antithrombin III — thrombin reaction is adopted as follows: heparin interacts with antithrombin III (the dissociation constant of heparin — antithrombin III complex is 10^{-6}–10^{-7} *M* [26]) which causes a conformational change of the antithrombin III macromolecule, thus promoting the reaction with the active center of thrombin [26–29], heparin being ultimately freed from the ternary heparin-antithrombin III-thrombin complex [30]. The antithrombin III-thrombin associate is a stoichiometric complex formed via the ester bond between the hydroxy group of the serine residue of thrombin and the carboxy function of asparagic or glutamic acid residues of antithrombin III. It is also admitted that the first step might be the direct interaction of heparin with the arginine residues of thrombin (the dissociation constant of the thrombien-heparin complex is 10^{-8}–10^{-9} *M*) occurs [31] which enhances the affinity of thrombin for antithrombin III [32]. An essential physiological role is assigned to the complexes of heparin with fibrinogen, thrombin, plasminogen, plasmin, and other compounds. These complexes were shown to exhibit pronounced fibrinolytic and anticoagulant activity [33,34]. Most of these are not only powerful antithrombins but also lytic agents as well (Table 1) [35].

Table 1 demonstrates that high lytic activity is displayed by the complexes of thrombin and fibrinogen as well as by that of heparin and plasmin. These complexes, as was shown in Ref. [35], affect the unstabilized fibrin only and supress simultaneously the activity of the fibrin-stabilizing factor. Qualitatively similar results were observed in *in vivo* experiments.

The anticoagulant activity of heparin depends also on its molecular mass. Most

Table 1. Antithrombinic and lytic properties of some heparin-protein complexes[35]

Substances	Thrombin time of oxalate filled plasma, (s)	The lysis time (h) at 37 °C of a clot produced from	
		plasma	fibrinogen
Heparin-thrombin	300	3	2.5
Heparin-fibrinogen	300	2	1.0
Heparin-fibrinolysin	300	4	3.0
	Control		
Heparin	240	Lysis is not completed in 4 h	
Fibrinogen	24	Lysis is not completed in 4 h	
Physiological solution	7	Lysis is not completed in 4 h	

researchers consider 10–12 repeating units with sulfated aminoglucose residues (which corresponds to a molecular mass of 7000–7500) necessary for the optimal activity of heparin [20]. Recently, pentasaccharide fragments of the heparin molecule were shown to correspond to the minimal sequence length providing association with antithrombin III and factor Xa [36]. There is an ambiguity concerning the effect of increasing molecular mass of heparin on its activity: according to Refs. [20,37], growth of the molecular mass of heparin (over 12 units) is not accompanied by any changes in its activity. Some of the data suggests a permanent increase of activity with increase in molecular mass, while according to others, the maximum activity corresponds to the molecular mass of 20000 [13]. The mechanism of "activity-molecular mass" dependence is not yet clear, although there is some data indicating that a high molecular specie has an additional site for binding antithrombin molecules [38]. Cooperative interactions between high-molecular heparin and thrombin and/or between the heparin-antithrombin complex and thrombin, which are not feasible for low-molecular mucopolysaccharides, are also possible.

3 Synthesis and Structure of Heparin-Containing Polymers

3.1 Ionic Immobilization

Nowadays the use of heparin for thromboresistance-enhancing surface modification of polymeric materials seem quite obvious. In fact, heparin seems to be destined for this purpose as of its nature, because it exhibits extremely high anticoagulant activity and posess enhanced stability in physiological media of the body. However, as is often the case with crucial discoveries, the ability of heparin to enhance thromboresistance of polymeric materials was found in 1961 almost by chance when examining the effect of electrochemical characteristics of a polymeric surface on the performance of the material when in contact with blood [11,39]. Such investigations were stimulated by the development of concepts concerning the structure of the internal surface of blood vessels, the structure of blood cells, and the possible mechanisms of the blood clotting process on natural and alien surfaces. It had already been established that the blood cells, including platelets, are negatively charged. The internal surface of blood vessels contains mucopolysaccharides, chondroitinsulfate, gialuronic acid, and heparitin sulfate, the sulfate and carboxylic groups of which endow a negative charge to the vessel surface [40,41]. The aforementioned was the basis for the studies of V. Gott which led to the discovery of the polymer thromboresistance-enhancing properties of heparin.

Beginning his research, V. Gott rested on the assumption that the repulsion of similarly charged platelets and the blood vessel wall plays an important role in the prevention of blood clotting at the natural vascular surface. The polymers were modified by coating them with graphite and subsequently by treating with benzalkonium chloride and heparin solutions. The first component of graphite-benzalkonium chloride-heparin polymers (GBCH), i.e., graphite, due to its high adsorption capacity is strongly attached to the polymer surface, and it keeps benzalkonium chloride tightly associated. Heparin is attached to benzalkonium chloride via ionic bonds which involve positively charged quaternary ammonium groups of benzalkonium chloride and negatively charged functions of heparin. The amount of benzalkonium chloride adsorbed is almost independent on the surface concentration of graphite and ranges from 2 to 4 ng/cm^2 on varying the amount of graphite from 200 to 1000 ng/cm^2. The amount of immobilized heparin does not exceed 1–3 ng/cm^2 [42,43].

GBCH-polymers were the first synthetic materials that displayed relatively high thromboresistance. For instance, poly(methyl methacrylate) grafts, having been coated with GBCH and implanted in dog's vena cava, were patent for 14 days, while uncoated PMMA grafts were totally covered with thrombin within the first 2 hours [44].

However, in spite of these promising results, the GBCH-based polymers did not gain wide practical application, and that for two major reasons: First of all, only rigid polymer plastics, e.g., polycarbonates and polymethyl methacrylate can be modified by this technique. Such widely used polymers — for endoprosthetic purposes — as organosilicone and other elastomers, poly(tetrafluoroethylene) etc., could not actually be subjected to graphitization. The second reason was of a much more crucial character: the stability of GBCH complexes in physiological media, blood in particular, was insufficient. The complexes when exposed to blood, were shown to dissociate washing heparin and benzalkonium chloride into the blood stream. The highest elution of heparin (up to 50%) was observed within the first 3 hours after implantation of the polymer [45], and after 35 days the heparin concentration on the surface had decreased by more than 90% [46]. The elution of basic quaternary amine into the blood stream caused a decrease in the surfacial concentration of heparin, causeing extensive hemolysis and damage to platelets [42,43].

Assuming that each heparin molecule corresponding to its size occupies the area of 2000 Å^2, it is easy to show that a monolayer of densely packed heparin molecules must be assigned the concentration of 0.1–0.2 ng/cm^2 [47]. This value is in agreement with the experimentally determined concentration of heparin (approx. 0.3 ng/cm^2) corresponding to a blood clotting time of 100 minutes [48]. Summarizing these data and taking into account that heparin coating is extremely non-uniform (the maximum and minimum coating thickness ratio reaches the value 30:1 [48]) one must admit that GBCH polymers are not suitable for the purpose of long-term implantation into the living body. During the contact with blood far one week, the heparin concentration is diminished to the extent where the surface of the polymer device is no more totally coated with heparin and displays regions that are free of heparin, thus loosing its thromboresistant properties. The inadequacy of the GBCH polymer-based implants was confirmed by the results of their clinical tests. During 1963–1965, supports for "petal" valves made from rigid GBCH polycarbonates were implanted to more than a hundred patients [49,50]. However, about 10% of the valves (the petals were made from unheparinized silicone rubber) were covered with thrombi, in the stagnation regions of the blood stream in particular. In 1965, these works were terminated.

On the other hand, the results generally seemed promising and gave impetus to further development in this field. This resulted in various procedures for making HCP with heparin attached to the polymer via either ionic or covalent bonds.

The studies of the stability of heparin-amine complexes demonstrated that the stability of this class of polymeric materials might be improved by choosing the right amine. Quaternary ammonium salts were shown to bind heparin stronger than primary, secondary, and tertiary amines, while amines containing alicyclic or aromatic fragments are more efficient than the aliphatic ones. Of the quaternary ammonium salts which have been examined, tridodecyl methyl ammonium chloride (TDMAC) —$(C_{12}H_{25})_3N^+(CH_3)Cl^-$ has become most widely used.

There are two methods for heparin coating of a polymer with the aid of TDMAC. The first one (a two-step procedure) involves treatment of a polymer with a TDMAC solution and an aqueous solution of heparin [51,52]. The treatment of the polymer with TDMAC solution is accompanied by its partial swelling (as the solvent is a 1:1 toluene/petroleum ether mixture) the hydrocarbon chain of TDMAC penetrates into the bulk of the polymer and stays there as if "stacked into" after evaporation of the solvent. Heparin is then bound to the quaternary ammonium groups localized on the surface. The second method (a one-step procedure) involves treatment of the polymer with a solution of heparin-TDMAC complex in toluene and subsequent elimination of toluene [53]. The time necessary for treatment depends on the solubility of the complex in the polymer. For organosilicone and vinyl polymers, simple immersion into the solution of the complex is sufficient to achieve adequate heparinization, while poly(tetrafluoroethylene), polypropylene, and high-density polyethylene require heating in the solution for several hours.

The TDMAC-based procedures make it possible to modify most of the commercially available polymers, heparin being much more strongly attached to the polymer than in the case when the graphite-benzalkonium chloride procedure is applied (Table 2).

The results of the *in vivo* tests of the so-modified polymers, although somewhat contradictory, are on the whole quite promising in respect to endoprosthetic implantation. Some investigators prefer "one-step procedure"-modified materials as no platelet adhesion and aggregation is observed while the plasma proteins and enzyme systems are not subject to any essential changes when the said materials are exposed to blood. The materials heparinized via a "two-step procedure", however, were also reported to be successful. For instance, the heparinized standard woven Dacron grafts when implanted into the dog's lower vena cava and into the abdominal aorta totally retained their patency for a period of 2 months, having been only covered with

Table 2. Immobilization of heparin using the procedure involving tridodecyl methyl ammonium chloride[53]

Polymeric matrix	Amount of immobilized heparin ng/cm²	The amount of heparin left at the polymer surface after treatment with	
		physiological solution, 100 h, %	plasma, 4 h, %
Poly(vinyl chloride)	25	100	95
Polycarbonate	4	93	90
Poly(ethylene terephthalate)	18	95	90
Silicone rubber	22	91	—
Polyurethane	186	100	98
Polyethylene	25	100	94
Polypropylene	8.2	92	91
Poly(tetrafluoroethylene)	1.1	99	87

a thin layer of neointima [54]. Relying on these results, a method for the preparation of heparinized shunts and vascular grafts of small diameter was developed. On the other hand, the arterio-venous shunt obtained by the "one-step procedure"-heparinized organosilicone rubber was reported [55] to be clotted within several hours to several days after being implanted between the carotid artery and the jugular vein of the he-sheep. This may be caused in part by the fact that heparin is rapidly washed off the polymeric surface by the highly turbulent and vigorous blood flow [55]. Studies with radioactively tagged heparin demonstrated that within a week after implantation, the surface concentration of heparin was decreased by 65–80%. To enhance the stability of HCP, implants were treated with glutaric dialdehyde which acted as a crosslinker for heparin molecules. The resulting enlarged heparin molecules multiplied in this way displayed lower elution rates as compared to the "initial" heparin molecules [56]. For instance, the concentration of "monomeric" heparin during the 3 hours of implantation was decreased by 75% while the crosslinked heparin was not actually washed off during a period of 12 hours [57,58]. It is by this procedure that the polymeric membranes for assisting oxygenators were modified [56]. The resultant membranes were stable for 48 hours.

An interesting procedure for heparinization of organo-silicone rubbers was suggested by Ui and Kammermeyer [59]. It involves the coating of a rubber-made product with a non-vulcanized organosilicone elastomer and subsequent incorporation of an anion-exchange resin into the latter simultaneously with vulcanization of the organosilicone layer, followed by introduction of a solution of heparin. The anion-exchange resins were: Dowex-1-X, Dowex 1-X10, Dowlite A1-101, and Amberlite XE-255. A modified organosilicone rubber sample was prepared as a blanck in which graphite was used instead of the anion-exchange resin; subsequent treatment was performed according to Gott.

The heparin content of the materials involving anion-exchange resins goes up to 800 ng/cm^2, whereas the maximal heparin content of graphite-based polymers is 0.78 ng/cm^2. All of the synthesized polymers, although exhibiting sufficiently high thromboresistance in *in vitro* tests (Table 3), drastically varied in their stability in various model media (Table 4).

As is seen from Table 4, heparin-containing polymers based on an organosilicone rubber matrix filled with anion-exchange resins demonstrate rather high stability in various physiological media. It should be noted, however, that the results of *in vivo* tests have not been very favorable — the thin-wall grafts made of heparinized organosilicone rubber implanted in dog's vena cava were thrombed in 2 hours after implantation. It is yet not clear why the *in vivo* thromboresistance is so low.

Table 3. The clotting time of human blood at some polymeric surfaces[59]

Polymeric material	Blood clotting time, min
Silicone rubber	20
Glass	4–6
Polymers heparinized by graphitization	> 480
Polymers heparinized via the procedures involving anion-exchange Amberlite XE-255 resin	> 480

Table 4. Stability of "heparin-organosilicon polymer" conjugates, treatment time — 24 hours[59]

Conjugating agent	Composition of solution used for the treatment	Heparin elution, %
Graphite	Distilled water	0
	0.9% NaCl solution	33.3
	Citrate plasma (human)	100
Amberlite XE-255	Distilled water	0
	0.9% NaCl solution	0
	5.0% NaCl solution	8
	95% ethanol	0
	Citrate plasma (human)	9.2
Dowlite A1-101	Distilled water	0
	0.9% NaCl solution	0
	5.0% NaCl solution	10.8
	95% ethanol	0
	Citrate plasma (human)	11.8
Dowex 1-X10	Distilled water	0
	0.9% NaCl solution	0
	5.0% NaCl solution	10.8
	Citrate plasma (human)	10

In summary, the studies of polymeric materials containing heparin ionically conjugated to a physically adsorbed layer onto the polymeric surface compound, give no reasons for being too optimistic. Although the majority of polymers tested *in vitro* appears to be thromboresistant, the results of *in vivo* tests are not that unambiguous. In spite of the fact that to date a number of devices has already been constructed on the basis of such HCP (e.g., artificial heart valves [60], blood filters [61], catheters, shunts, some parts of the artificial heart and kidney, membranes, etc. [56], some of which are even used in clinics, most investigators tend not to overestimate these materials regarding their long-term implantation in a living body. This pessimistic viewpoint is justified when taking into account the low stability of HCP in physiologically active fluids, vigorously flowing ones in particular. When heparin or its complex with the corresponding cation is washed off, the uncoated heparin-free surface regions trigger thrombus formation.

The above supposition [62] is confirmed by the studies of thromboresistance of polytetrafluoroethylene heparinized in a "two-step TDMAC procedure" as a function of the size of uncoated regions. The heparin coating was eliminated by one of the following procedures: 1) placing the material in toluene; 2) ultrasonic treatment; 3) mechanical treatment damaging the coating. In all cases heparin was eliminated as a complex with TDMAC. The results of these studies, compiled in Table 5, reveal the total loss of thromboresistant properties of the polymer when the diameter of the uncoated regions exceeds 20 mkm. This is usually observed when the coating thickness is reduced by 30% on the average.

Better prospects for the synthesis of HCP for long-term use might have been anticipated with the introduction of the heparinization procedure involving covalent

Table 5. Thromboresistance of heparinized poly(tetrafluoroethylene) as a function of the diameter of uncoated regions[62)]

Method for elimination of coating	Average thickness of the coating, Å	Diameter of uncoated regions, mkm	Clotting of plasma[a] occurs in		
			40 min	80 min	120 min
Mechanical degradation	3500	1–2	0	0	0
	3000	1–2	0	0	0
	2500	1–2	0	0	0
	2000	2–3	0	0	0
	1500	10	0	0	+
	1000	10–20	0	0	+
	500	25–50	0	0	+
	100	> 50	+	+	+
Ultrasound	3500	1–2	0	0	0
	3000	1–2	0	0	0
	2500	1–2	0	0	0
	2000	2–3	0	0	0
	1500	10	0	0	+
	1000	10–20	0	+	+
	500	25–50	0	+	+
	100	> 50	+	+	+
Treatment with toluene	3500	1–2	0	0	0
	3000	5–10	0	0	+
	2500	10	0	0	+
	2000	10	0	+	+
	1500	10–20	0	+	+
	1000	25	0	+	+
	500	> 50	+	+	+
	100	> 50	+	+	+
Initial polymer	0	—	+	+	+

[a] 0: no clot formed; +: Clot formed

attachment of the positively charged groups onto the polymer surface and subsequent ionic immobilization of heparin on this surface.

One of the pioneering works in this field was the heparinization of polystyrene via its chloromethylation followed by quaternization with dimethylaniline and coupling with heparin [63)]:

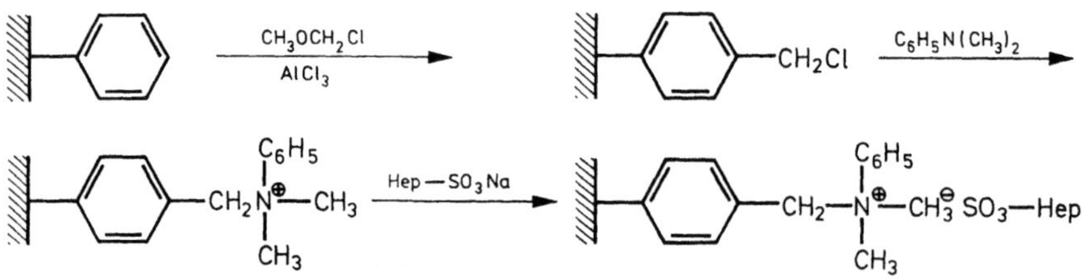

Polystyrene itself is not used for endoprosthetic purposes and its application is accounted for only because of easy substitutions in benzene rings. The method was subsequently modified for heparinization of silicone and natural rubber, polyethylene, polypropylene, poly(ethylene terephthalate), and other polymers. Styrene was first grafted onto the polymers by γ-radiation and then the above-described reaction was performed in the second step. All the polymers synthesized in this way contained sufficiently large amounts of immobilized heparin (2.8—15.7 ng/cm^2) and displayed good thromboresistance when tested *in vitro* — recalcified blood was not clotted for several hours.

Chemical groups capable of ionic binding of heparin may be introduced onto the polymer surface by radiation grafting of ionogenic monomers, e.g., 2- or 4-vinylpyridine, dimethylaminoethyl methacrylate, etc. [64,65]. Heparinization proceeds as shown below:

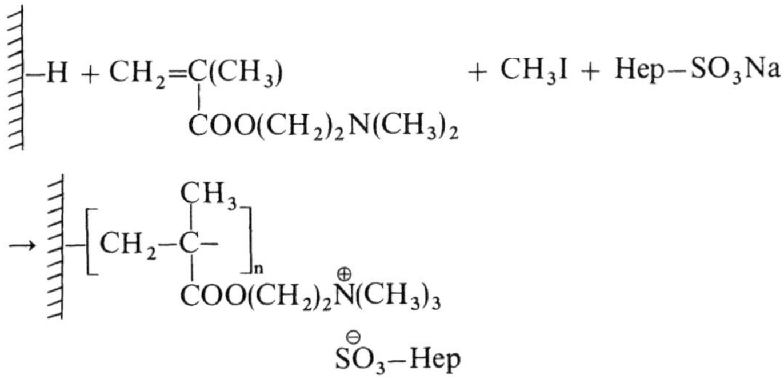

The stability of heparin-amine complexes depends to a large extent on the structure of the constituent amine. The best results were obtained for dimethylaniline and pyridine (Table 6) [65]. The amount of bound heparin is, as is seen, proportional to the amount of immobilized amine, while the molar ratio amine/heparin is predetermined by the structure of the surface. Microscopically rough surfaces were shown to prevent association between heparin and amine as the portion of the amine is sterically inaccessible for the macromolecules of heparin.

Table 6. Amount of heparin bound as a function of the nature and amount of covalently immobilized amine[65]

Surface	Alkylated dimethyl aniline, 10^9 mole/cm^2	Alkylated pyridine, 10^9 mole/cm^2	Heparin content, 10^9 mole/cm^2
Graft copolymer of styrene	6.51	—	0.607
and polypropylene	—	2.57	0.294
Silicone rubber	15.2	—	3.03
	—	18.5	2.82
Graft copolymer of styrene	55.7	—	4.74
and silicone rubber	—	59.4	3.84

Table 7. Elution of heparin from the surface of polypropylene with the solutions of plasma proteins, 25 °C, time of incubation 30 min[65]

Protein	Concentration of protein in solution, mg/ml	Heparin elution, %
γ-Globulin	10.0	54
α-Globulin	4.0	92
β-Globulin	4.0	56
Fibrinogen	3.6	64
Hageman factor	0.01	64
Albumin	20.0	59

Amino groups may be grafted to the polymer surface by treatment with low-temperature gas plasma (the initial gas may either be ammonia or a mixture of nitrogen and hydrogen) [66]. Various polymers, including polypropylene, polyvinyl chloride, poly(tetrafluoroethylene), polymethyl methacrylate, polyurethanes, and polycarbonates, etc. were heparinized using this procedure. The polymer layer subjected to modification in glow discharge is less than 1000 Å, which enables to retainment of the mechanical properties of the initial polymer.

The stability of such HCP is not high. For instance, although the heparin (ionic bonds) — ternary ammonium salts (covalent bonds) — polypropylene associate is not actually affected by distilled water, the 3 hours storage of the product in blood plasma reduces the surface concentration of heparin to 30% of its initial value [65]. It is noteworthy that heparin is very effectively eluted by solutions of γ-globuline (Table 7) [65].

One of the prospective pathways for the synthesis of HCP involves copolymerization of neutral monomers with a basic ionogenic one. This procedure was used for synthesis of copolymers of acrylonitrile and dimethylaminoethyl methacrylate used for production of dialysis membranes [67]. Prior to use, the membranes were quaternized with HCl and subsequently treated with heparin. The reported high stability of the heparin-polymer complex (the loss of heparin on washing the polymer with distilled water for 70 hours was less than 1%) does not obviously ensure that the HCP will perform properly when contacted with solutions of proteins and blood.

Quite promising results were obtained when a hydrophilic monomer and a monomer containing a basic group were simultaneously grafted onto a polymer. This seems to be rather sensible as, together with the incorporation of the groups interacting with heparin, hydrophilization of a hydrophobic polymer surface is achieved. Whereas hydrophilization by itself results in the enhancement of thromboresistance, as is emphasized by many investigators, the hydrophilic environment ensures a better medium for the functions of immobilized heparin to be performed.

As an example of the above approach, let us review the rsults of series of investigations by Japanese researchers [68–75]. HCP was synthesized by treating the copolymer containing 56 mole% of vinylchloride, 12 mole% of vinyl acetate, and 32 mole% ethylene (Graftmer R_3) with N,N-diethyldithiocarbamate. The thus introduced dithiocarbamate groups are highly efficient in co-initiating graft-copolymerization.

Graft-copolymerization of poly(ethylene glycol monomethacrylate) of the formula:

$$CH_2{=}C(CH_3)\text{—}CO(OCH_2CH_2)_3OCH_3$$

and dimethylaminoethyl methacrylate was subsequently effected by UV irradiation, the substituted amino groups were quaternized with ethyl bromide, and the resultant polymer was treated with heparin. The heparin content was independent of the composition of the grafted monomer mixture and was about 14–19 wt.-%, while the water uptake of the swollen copolymer was proportional to the content of the hydrophilic monomer. The stability of HCP is a function of the water content in the swollen copolymer (Fig. 1). As is seen in Fig. 1, the higher the water content, the easier was the heparin eluted. Considering that the elution rate is a function of the dissociation constant of the heparin-amine complex and of the diffusion of heparin in a layer of grafted copolymer, the observed dependence may easily be explained by a better diffusion of heparin in a polymer with a high water content.

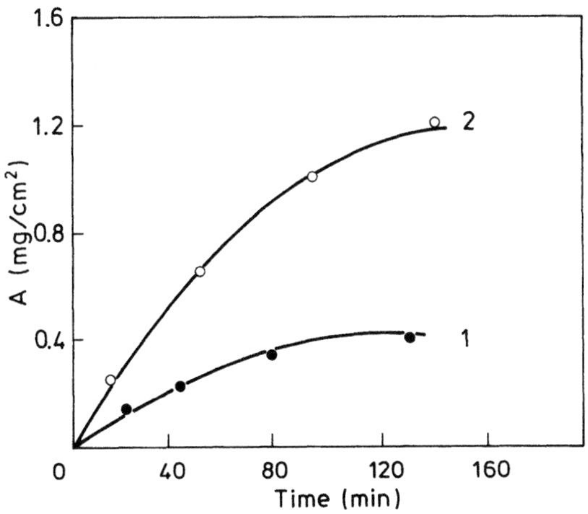

Fig. 1. Time dependences of the elution of heparin with a 1.0 *M* NaCl solution at 37 °C from the surface of polyurethane catheters coated with the copolymer containing: 14% heparin and 18% water (1), and 17% heparin and 44% water (2) [68]

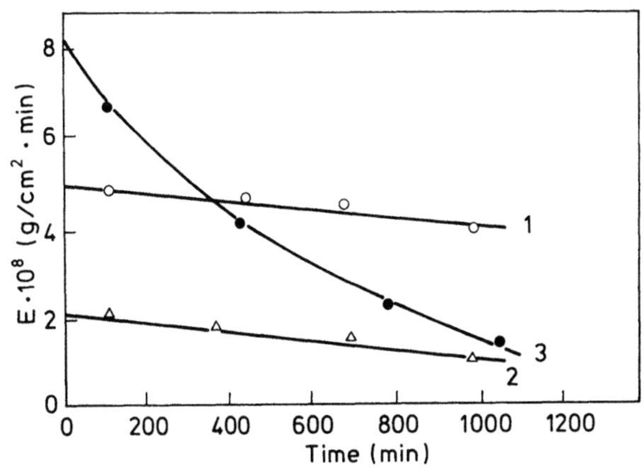

Fig. 2. Rate of heparin elution with the blood plasma out of the grafted copolymers containing 15% heparin and 29% water (1), 17% heparin and 12% water (2), 11% heparin and 36% water (3) [71]

Table 8. Results of clinical tests of catheters implanted for periods of time ranging from 2 to 85 days[70]

Material of the catheter	Number of implantations	Number of thromboses
Poly(tetrafluoroethylene)	2	2
Silicone rubber	8	5
Poly(vinyl chloride)	4	4
Polyethylene	4	4
Polyurethane coated with a copolymer containing 17% heparin and 44% water	81	9

All the synthesized HC copolymers were negatively charged, as only the part of the negatively charged groups of the heparin macromolecule is involved in the reaction with the quaternized ammonium groups, while the other remain unneutralized.

Tests *in vivo* give best results for copolymers with 30% water content. The specific feature of their performance, as compared to the copolymers of other compositions, is a relatively high and constant rate of heparin elution (Fig. 2).

The authors of the above-discussed results consider the constant rate of heparin elution to be absolutely necessary for a successful functioning of the implant. Catheters made of these copolymers are much better, in respect of their thromboresistant properties, than the commercially available poly(tetrafluoroethylene) polyethylene, and plasticized polyvinyl chloride catheters (Table 8). The tested catheters were clotted in 9 cases of 81 (11%), whereas usual silicone rubber catheters were clotted in 5 cases of 8 (63%) [70].

In cases when graft-copolymerization is, for any reasons, difficult to carry out, a procedure may come to be effective which was developed for heparinization of silicone rubber [76].

Graftcopolymerization onto silicone rubber is rather difficult to achieve and is often accompanied by unwanted changes in physico-mechanical properties of the polymer caused by initiating agents. To overcome the problem, silica was introduced into the rubber matrix as an active filler capable of binding cationic compounds; such a cationic compound being γ-aminopropyltriethoxysilane. Schematically the pathway for heparinization of the latter may be presented as follows:

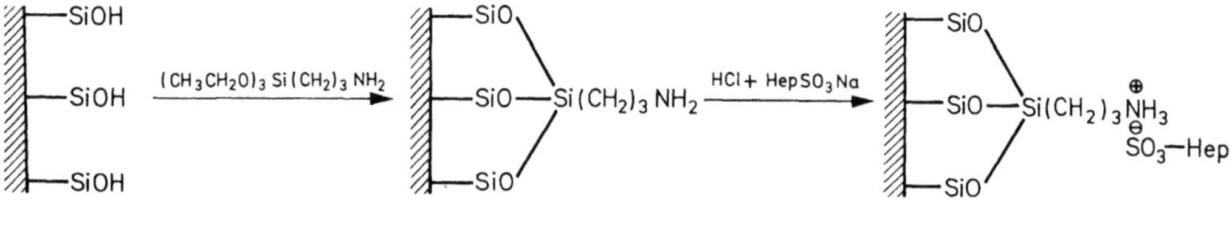

A similar procedure was reported for heparinization of glass [77].

The *in vivo* tests did not give any satisfying answer to the problem of long-term contact of heparin-silicone rubber with blood. The blood-surface interaction was essentially dependent on the procedure adopted for sterilization of the polymeric material. For instance, there was no substantial platelet adhesion on the polymer subjected to sterilization with ethylene oxide. The minor amount bound was neither activated nor aggregated. Sterilization in an autoclave with steam resulted in a significant increase of the adhesion and activation of the platelets [78]. These results demonstrate the importance of the proper choice of a sterilization procedure as well as the necessity to consider all parameters when undertaking comparative studies using a HCP.

There are several other methods for ionic attachment of heparin to polymeric materials worth mentioning. Dimethylaminoethylcellulose was treated with heparin [79]. The products made from a blend of this polymer with organosilicone polymers were still in good condition after a period of more than a year. The *in vitro* blood clotting time was over 24 hours.

Grafting ethyleneimine onto cellophane followed by conventional heparinization yielded a product used in making the membranes for artificial kidney which displayed good thromboresistance [80].

The free amino groups of a polymer, poly-*p*-aminostyrene for instance quaternized with methyl iodide are also capable of binding heparin [81].

A similar procedure was used for the heparinization of block copolymers composed of the poly-adduct of N,N′-dimethyl-ethylenediamine and 1,4-bisacryloylpiperidine as a central block and of polystyrene as the end blocks [82]. To enhance the stability of the material, heparin was crosslinked with glutaric dialdehyde. The resultant HCP sustained a 6 month implantation into the dog's vena cava.

In the works of Ferruti et al. [83,84], there has been reported a method for heparinization of plasticized polyvinyl chloride which was pre-modified by grafting of polyaminoamines. The surface concentration of ionically bound heparin was 1.2 ng/cm^2. Heparin can be eluted off the polymer in a narrow pH range of 10.8 to 11.4.

In summary, it must be noted that in spite of the wide-spread utility of the methods of ionic immobilization of heparin it is difficult to consider a long-term use of the resultant HCP. All of the reported results indicate that after the contact of HCPs with blood and other physiological media, heparin is quite rapidly desorbed from the polymeric surface. This does not restrict the use of such heparin-containing polymers for the purposes of a short-term contact with blood. Today such materials are already used to manufacture catheters for perito-venous shunting in Japan, USA, France, and Sweden [85–87].

3.2 Covalent Immobilization

Heparin is most tightly bound to a polymeric material when immobilized via covalent bonding. The heparin molecules contain hydroxy, carboxy, sulfate, and amino groups which are all suitable for this purpose. However, heparin is soluble in water and formamide only and is insoluble in organic solvents which presents a significant restriction for covalent conjugation of heparin.

One of the pioneering works in this field was the covalent conjugation of heparin to poly-*p*-styrene isocyanate [81]. The procedure involved the following steps:

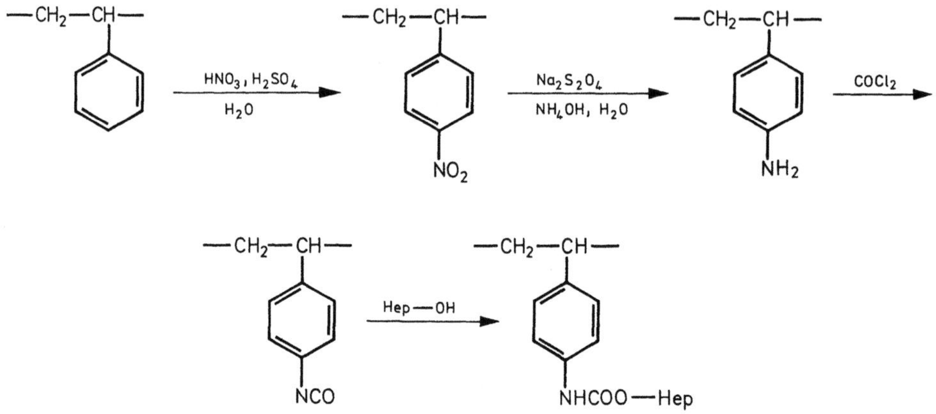

The thromboresistant properties of polystyrene with ionically immobilized heparin is unambiguously inferior compared to the covalently bound one. For instance, the blood clotting time for some of the "covalently immobilized heparincontaining samples" was up to 720 min, while for polystyrene, poly-*p*-aminostyrene, and poly-trimethyl aminostyrene with electrostatically bound heparin the blood clotting time did not exceed 35 min.

The conjugation agents used for covalent immobilization of heparin were hydrazide, ethyleneimine, carbodiimide, etc. The best results were obtained when heparin was immobilized via its intermediate adduct with cyanurchloride. The treatment of heparin with cyanurchloride produced a heparin derivative containing 7 cyanurchloride residues per heparin molecule [64], whose anticoagulant activity was 85% of the activity of the initial unmodified heparin. To immobilize the adduct on silicone rubber, the latter was treated with γ-aminopropyltriethoxysilane. The alternative procedure involved the reaction of cyanurchloride, di-acid chloride, or diisocyanate with the heparin "portion" of the heparin-TDMAC complex [4]. The complex was soluble in organic solvents, which made it possible to use a number of water-insoluble conjugating agents. The adduct produced was then reacted with functional groups of the matrix polymer. In both procedures, the heparin content reached up to 500 ng/cm^2. Obviously, the bonding strength of attachment of heparin to the polymer is much greater for the covalent immobilization than for the ionic binding of heparin (the elution of covalently and ionically immobilized heparin was 3–4% and 45%, respectively).

It seemed reasonable to anticipate that the synergism of these two features (high heparin content and stability of the resultant materials) would result in long-term thromboresistant polymers. The *in vivo* tests revealed, however, their extremely low thromboresistance as compared to the "ionically bound heparin-containing" polymers, in particular. The effect is assumed to be caused by a lack of sufficient mobility of the polymer-bound heparin molecules, which prevents the performance of the intrinsic anticoagulant properties of heparin.

Another procedure for covalent immobilization of heparin involves its binding to a poly(vinyl alcohol) matrix effected by the mixture of formaldehyde and glutaric dialdehyde [88,89]. The films produced from the solutions of this polymer displayed good thromboresistance but poor mechanical strength. They could thus be applied only with a reinforcing network. Stronger materials were obtained by dehydration at 80 °C for 100 min of the solution containing poly(vinyl alcohol), formaldehyde, glutaric dialdehyde, glycerol, magnesium chloride, and sodium heparinate. The resultant material was subsequently hydrated [4].

Polystyrene-polybutadiene-polystyrene block-copolymers were also heparinized via formation of acetal bonds. The procedure involved oxidation of the copolymer to glycol [90]:

$$-CH{=}CH- \xrightarrow[\text{acid}]{\text{peracetic}} -\underset{\diagdown\;O\;\diagup}{CH-CH}- \xrightarrow[H^+]{H_2O} -\underset{OH}{\underset{|}{C}H}-\underset{OH}{\underset{|}{C}H}-$$

Hydroxyl groups may also be introduced in the polymer by graft-copolymerization of hydroxyethyl methacrylate or glycidyl methacrylate, the latter being subsequently hydrolyzed [91]. Heparin was immobilized on such hydroxyl-containing polymers by means of glutaric dialdehyde. The maximal amount of polymer-bound heparin was 80 ng/cm^2.

A perspective method for the production of HCP was proposed by Labarre et al. [92–94]. It involves radical graft-copolymerization of methyl methacrylate onto heparin effected by cerium salts (Ce^{4+}). As well as other hydroxyl-containing polymers, heparin may react with cerium ions producing free radicals which may initiate graft-copolymerization of unsaturated compounds:

$$\text{Heparin}-OH \xrightarrow{Ce^{4+}} \text{Heparin}-O^{\cdot} \xrightarrow{CH_2=\underset{X}{\underset{|}{C}H}}$$

$$\rightarrow \text{Heparin}-O-(CH_2-\underset{X}{\underset{|}{C}H})_n-$$

The prospects for this method as well as for the method involving the generation of free macroradicals by γ-irradiation of heparin [95], are provided for by the variety of polymerizable monomers, which makes it possible to produce materials with various physico-mechanical and chemical properties. The *in vitro* thromboresistance of the copolymers obtained in this way was proportional to the heparin content (Table 9).

The main reason for the increase of the clotting time is provided by the covalently immobilized heparin but not by the eluted one. This is illustrated by Fig. 3, showing the plasma clotting time *vs* the amount of immobilized heparin or leaking into plasma.

One can see (Fig. 3) that the amount of heparin eluted, if eluted at all, is so small that it cannot account for the clotting time observed. After the copolymers are exposed to plasma, their activity decreases and may be restored only by special treatment.

Heparin immobilized using the described procedure retained only 1 % of the initial activity of heparin in solution. This fact may probably be explained by the denaturation of heparin effected by nitric acid, which was the reaction medium for the graft-copolymerization procedure. The inaccessibility of the majority of heparin macromolecules of a copolymer not swelling in an aqueous solution, in particular, supplies

Table 9. Thromboresistance of heparin-(methyl methacrylate) copolymers* [93]

Heparin content, %	Thrombin time, min	Recalcification time, min
0	0.51	2.70
1	2.47	2.67
2	2.67	13.00
4.5	6.00	34.00
11	60.00	60.00

* Determined as follows: 1 mg of the copolymer was dispersed in a 0.1 ml solution of non-ionogenic surface-active compound at 37 °C for 10 min; 0.2 ml of plasma were added to the suspension; the mixture was incubated for 15 min at 37 °C; 0.1 ml thrombin (5 U/ml) (thrombin time) or 0.2 ml of 0.025 *M* $CaCl_2$ (recalcification time) were then added and the time of clot formation was measured.

another reason for the observed loss in activity. Besides this, the decrease in activity may be attributed to the changes in the chemical composition of heparin due to the partial changes in the nature of its functional groups, particularly the hydroxyl groups in the discussed case.

The contribution of the above given parameters to the total affect may be monitored by comparing the results of immobilization of heparin on hydrophilic gels using various conjugating agents. On the dextrane gels which are most common, heparin may be immobilized as via free amino groups, so as via carboxy groups [81].

Summarizing the data reported in the literature, one may conclude that the activity of immobilized heparin does not depend on the nature of the functional group engaged in the immobilization (OH or COOH), especially when the concentration of the blocked groups is relatively small.

Thus, the activity of immobilized heparin is effected most by the mobility of its macromolecule and its accessibility for physiological substrates [96,97]. These parameters may easily be controlled by changing the structure of the hydrophilic gel and

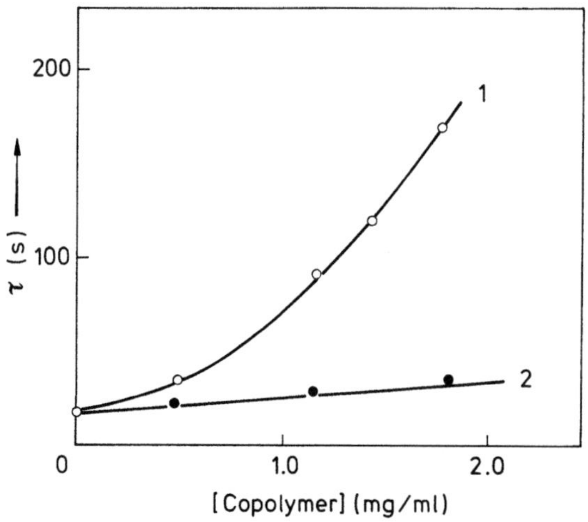

Fig. 3. Plasma clotting time as a function of amount of copolymer in contact. 1 — 0.1 ml of a suspension of copolymer was incubated with 0.2 ml plasma for 20 min, 0.1 ml thrombin solution was added; 2 — 0.1 ml of the copolymer suspension was incubated for 20 min with 0.2 ml plasma, the copolymer filtered from the solution, and to this solution were then added 0.1 ml buffer solution and thrombin solution [93]

Table 10. Activity of immobilized heparin as a function of the immobilization procedure [81]

Matrix	Conjugating agent	Amount of immobilized heparin mg/g polymer	Amount of heparin cofactor bound per 1 g of product, antithrombin units
Agarose	cyanobromide	5	< 2000
Tetramethylene-diaminoagarose	thiophosgene	60	30000
ε-Aminocaproylagarose	carbodiimide	60	30000

varying the length of the spacer between the heparin molecule and polymeric backbone (Table 10) [81].

The use of the spacer is seen to lead to the increase in the amount of immobilized heparin and to a less dramatic decrease of the activity of heparin due to immobilization (Table 10), although the activity of the immobilized heparin is still below 10% of the initial activity of the free heparin [81].

A rather similar dependence was observed by Hoffman et al. [98] for heparin covalently immobilized on silicone rubber, which had been pre-modified by graft-copolymerization of hydroxyethyl methacrylate and N-vinylpyrrolidone (Table 11).

Table 11 demonstrates, as in the previous case, the increased amount of immobilized heparin when the spacer was used; the anticoagulant activity of immobilized heparin being around 10% of the initial heparin activity. The reason for the observed loss of initial activity is probably the inaccessibility of the immobilized heparin for high-molecular proteins (thrombin with a molecular mass of 34000 and antithrombin III with a molecular mass of 65000) as the permeability of the grafted gels, whose water content was only 55%, may be insufficient.

At the same time there are cases reported where the amount of immobilized heparin

Table 11. Amount of immobilized heparin as a function of the method of immobilization [98]

Method of immobilization	Type of the bond	Amount of heparin, mg/cm^2
Initial polymer	physical	1.32
Cyanobromide	covalent	2.06
Cyanobromide + ε-aminocaproic acid	physical	2.22
Cyanobromide + ε-aminocaproic acid + carbodiimide	physical	0.44
Cyanobromide + ε-aminocaproic acid + carbodiimide	covalent	12.60

was actually independent of the spacer length, although the anticoagulant activity of the modified polymers was shown to increase with the increase in the spacer length [21,99]. This is true for covalent immobilization of heparin onto polyurethane matrices, carbodiimide being the conjugating agent, and for immobilization of heparin on diaminoalkylagarose by means of the Woodward reagent. Polyurethanes were partially hydrolyzed to obtain free amino groups on the surface. These were transformed to isocyanate groups by phosgene treatment, the diaminoalkane spacer was incorporated and heparin was immobilized by the carbodiimide procedure.

In all of the reported cases, anticoagulant activity of immobilized heparin, as was already stressed above, was less than the activity of heparin in solution, which was attributed to the low accessibility of heparin molecules for physiological substrates. Increasing the porosity of the polymer, i.e., increasing the accessibility of the immobilized heparin led to an increase in its anticoagulant activity [100].

A method for *in situ* immobilization of heparin in a highly swollen matrix of a synthetic polymer was proposed which ensured the high mobility of heparin molecules, the absence of steric hindrances for its interaction with high-molecular substrates, and strong attachment of heparin to the polymer matrix [101,102]. In the described procedure the polymeric matrix is formed in the presence of heparin, which ensures the preservation of the heparin molecules of their initial conformational state and anticoagulant activity. The strength of heparin-polymer bonding is also sufficient, as is demonstrated in Fig. 4: a fraction of radioactively tagged heparin was washed out from the gel by blood plasma with a constant rate of radioactivity desorption from the gel. The slight washing out of heparin by a physiological solution and by blood plasma from the gel pre-washed with water, results from the fact that upon changing the water for these solvents, the structure of the gel is changed (contraction of the polymer network), and unbound heparin, left after washing with water, is then driven out of the gel into solution. The radioactivity of the plasma-washed gels remained constant for 300 days and by no known analytical method could heparin be detected in waste waters.

A similar approach was used in Ref. [103] to obtain the fat-soluble derivatives of heparin by its copolymerization with butyl methacrylate or vinyllaurate and to get high-molecular products with a molecular mass of over 200000 by homopolymerization of the unsaturated heparin derivative. The products were used as thromboresistant

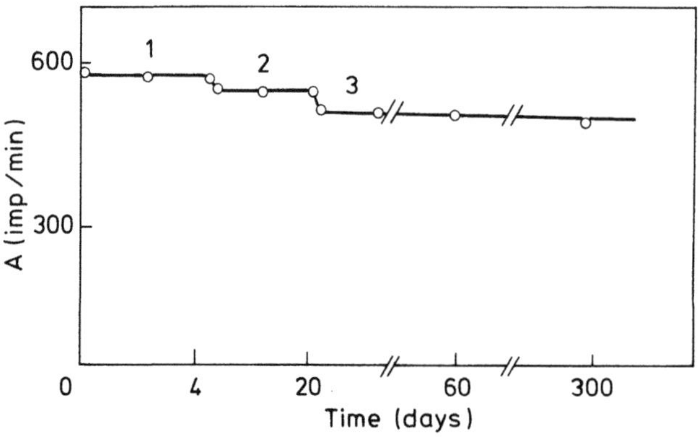

Fig. 4. Radioactivity of the ^{14}C-labelled heparin-containing hydrogel as a function of eluent and elution time. 1 distilled water, 2 physiological solution, 3 human blood plasma [101]

coatings for various materials. The unsaturated derivative introduced into polymerization was synthesized by acylation of heparin with methacryloyl chloride.

Summarizing the experimental data of this Section one may come to the following conclusion. The methods of ionic immobilization of heparin today have become the most widely used, because of the simplicity of these methods and the possibility to modify a large number of polymeric material, having relatively high thromboresistance being ultimately attained. The utility of these materials is restricted by a rapid washing-off of heparin from the polymeric surface, although a number of successful implantations lasting up to several months were reported.

Heparin is actually not washed out at all or is eluted at a very low rate when it is covalently bonded to the polymeric material. However, quite a drastic decrease of anticoagulant activity of heparin is often observed in that case.

The fact that thromboresistance of HCP is dependent on the method of immobilization of heparin, together with a rather low activity of covalently immobilized heparin, makes the idea of long-term enhancement of thromboresistance of polymers on their heparinization doubtful [54,70,71]. Naturally, the answer can be given only after a detailed analysis of the interaction of HCP with blood and its components and clarification of the mechanism of the effect of the immobilized heparin on the blood clotting system and relying on the results of *in vivo* tests of these materials are necessary.

4 Interaction of Heparin-Containing Polymers with Blood and Its Components

A variety of methods for the synthesis of HCP has made it necessary for investigators to choose the one providing optimal thromboresistant properties for a given material with respect to certain working conditions. The right choice of the modification procedure, based on the prognosis of the performance of a HCP in a living body, may be performed only on the basis of the knowledge of the mechanism of antithrombogenic action of the immobilized heparin. However, this problem is not yet clear.

We shall introduce the discussion of the hypotheses of the mechanism of the phenomenon with a survey of the stages of the clotting process that may be affected by the polymer-bound heparin.

At present, the following scheme of the clotting process triggered at any alien surface, polymeric surfaces in particular, is accepted to hold true: The first step of the interaction of blood with a polymer is the adsorption of proteins, which may be either favorable or unfavorable for the subsequent cascade of reactions resulting in thrombus formation. For instance, predominant adsorption of glycoproteins, e.g., fibrinogen or γ-globulin, leads to increased platelet adhesion, whereas the adsorption of albumin "passivates" the polymer surface towards platelets. The negative role of glycoproteins in this case is determined by the constituent oligosaccharide fragments which may interact with some of the enzymes localized in platelet membranes (e.g., glycosyl transferase). Platelet adhesion constitutes the second step of the blood-polymer interaction. The amount of adhered platelets is governed, on the one hand, by the composition of the adsorbed protein layer, and, on the other hand, by the degree of activation of the contact factors of the blood clotting process, viz. factors XII (the Hageman factor) and XI (plasma precursor of thromboplastin). The activated contact factors initiate a cascade of successive enzymatic reactions resulting in the formation of the free thrombin which stimulates additional adhesion and aggregation of the platelets. Platelet adhesion is almost always accompanied by their activation and evolution into the surrounding medium of the intracellular granule contents (mainly serotonine and adenosine diphosphate), which in its turn accelerates platelet adhesion. The ultimate culminating step involves specific thrombin proteolysis of fibrinogen to fibrin-monomer, whose subsequent polymerization results in the formation at a polymer surface of the insoluble fibrin clot. The formation of the thrombus is actually completed when the cellular blood elements are entrapped between the fibrin strands.

Although the scheme presented does not reflect all details of the blood-polymer interaction, it is possible to evaluate the effect of the immobilized heparin on the above or other steps of the thrombosis process.

In general, the effect of immobilized heparin on the blood contacting a polymeric surface consists in an increase of the blood clotting time even when no heparin is eluted at all (Table 12) [104–108]

Heparinization, as is seen from Table 12, results not only in the increase in blood clotting time, but also in an increase of the platelet adhesion; the increase in both cases being more intensive the higher the content of the immobilized heparin. It should be noted that there is a complete agreement between the scientists concerning the effect of heparin on the blood clotting time: the result of the contact of blood with HCP is considered analogous to that of the direct injection of soluble heparin into the bloodstream, whereas the data on platelet adhesion is quite contradictory.

For instance, the modification of silicone rubbers, poly-(tetrafluoroethylene), polyethylene, and segmented polyurethanes by ionic attachment of heparin was shown to result in a decrease of platelet adhesion onto these polymers [71,74,75]. A similar

Table 12. Thromboresistant properties of some polymers containing covalently immobilized heparin [107]

Polymeric matrix	Concentration of bound heparin, mg/cm^2	RBCT[a] ± 10%	RIPA[b] ± 20%
Polyethylene	0	1.2	1.8
Polyethylene	0.3	1.8	3.9
Polyethylene	1.5	10	6.0
Polyurethane, USSR	0	1.7	1.0
Polyurethane, USSR	1.7	12	4.0
Polyurethane, USSR	2.9	>12	7.0
Polyurethane, CSSR	0	1.5	1.0
Polyurethane, CSSR	1.1	12	3.0
Polyurethane, CSSR	1.4	>12	4.0
Biomer, USA	0	1.7	0.5
Biomer, USA	1.5	10	6.0

[a] RBCT (relative blood clotting time) — the time of blood clot formation on a polymer surface/the time of blood clot formation on a glass surface;

[b] RIPA (relative index of platelet adhesion) — number of platelets adhered by a unit surface of polymer/number of platelets adhered by a unit standard surface (glass)

Table 13. Number of adhered platelets as a function of the aorta treatment procedure [110]

Modification of the aorta surface with	Number of platelets adhered per 1 mm^2
Initial, non-modified	199000
Heparin	14000
Thrombin	471000
Thrombin succeeded by heparin	44200

effect was reported in Ref. [109] on heparinization of polyethylene. Besides, the inhibitory effect of heparin on platelet adhesion may be displayed directly as well as via inhibition of thrombin [110]. This effect is illustrated in Table 13.

There is further data confirming the results compiled in Table 12. Figure 5 shows the results of the studies on the interaction of heparinized silicone rubber with blood [4]. The surface is seen to adhere the platelets up to its complete saturation, which then prevents further interaction.

Such an effect of increased platelet adhesion on heparinized materials was also reported in Refs. [88,111,112].

Quite significant is the fact that the amount of platelets adhered depends on the synthesis procedure for HCP. For example, the platelet adhesion onto silicone rubbers heparinized via the TDMAC procedure was 150000 platelets/cm^2, while the very same rubbers that were heparinized via γ-aminopropyltriethoxysilane adhered only 90000 platelets/cm^2 [88]. The platelets are to a greater extent adhered by the polymers containing covalently immobilized heparin than by those that elute heparin into the bloodstream [113] although the immobilized heparin itself does not interact with the platelets [21].

The diversions concerning the effect of heparin on platelet adhesion may probably be attributed to the different "binding strengths" of heparin to the polymer. When heparin is, at least to some extent, leaking into the bloodstream, the platelet adhesion

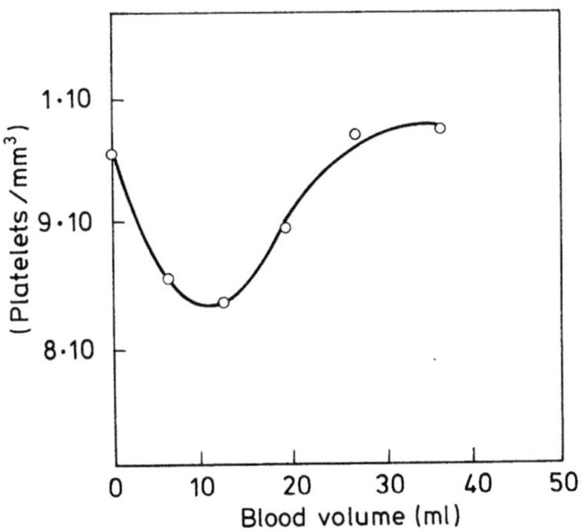

Fig. 5. Platelet count of blood having been contacted with heparinized silicone rubber as a function of blood volume [4]

is usually decreased, and on the contrary, strong attachment of heparin to a polymer increases the platelet adhesion.

However, in spite of the significant diversity in the quantitative evaluation of platelet adhesion, the platelets adhered onto heparinized surfaces are neither aggregated nor activated [4,78]. In other words, the intracellular contents of the adhered platelets, which may otherwise contribute to the blood clotting process, is not evolved into the bloodstream. This conclusion is verified by the results of Table 12, which indicate that the increased platelet adhesion does not result in a worse thromboresistance: the blood clotting time at a HCP surface is an order of magnitude higher than that at the surface of the initial non-heparinized polymer.

The reasons for the increased platelet adhesion onto covalently-immobilized heparin-containing polymers were disclosed in the studies of the protein adsorption onto HCP i.e., as stated above, the first step of the complicated blood-polymer interaction.

The data on protein adsorption as affected by immobilized heparin is quite contradictory. In Refs. [114] and [115], serum albumin was shown to be predominantly adsorbed by the HCP, whereas in the majority of other works the observed dependence was the opposite. Table 14 compiles the results of the studies of the absorption of plasma proteins by HCP of two different types [64]. Such thrombogenic proteins as fibrinogen and thrombin are seen to be the ones adsorbed most.

The polymers containing covalently immobilized heparin also display increased adsorption of fibrinogen and thrombin [94,106–108]. For instance, the maximal amounts of fibrinogen, thrombin, plasmin, and albumin adsorbed by the "covalently immobilized heparin"-containing hydrogel are 1.54, 0.59, 0.02, and 0.01 mg per 1 mg of immobilized heparin, respectively [106,107]. The predominant adsorption of fibrinogen as compared to that of serum albumin, was confirmed in quantitative studies of the adsorption of ^{125}I-labelled fibrinogen and ^{131}I-labelled albumin onto polyethylene having a covalently-immobilized-heparin content of 0.01 mg/cm^2. While predominantly albumin was adsorbed by the initial non-modified polyethylene (the concentrations of adsorbed albumin and fibrinogen after a 30 seconds exposition were 0.8 + 0.1 mg/cm^2 and 0.2 + 0.05 mg/cm^2, respectively), its heparinization reversed the mode of adsorption to the opposite and the concentrations of adsorbed albumin and

Table 14. Protein adsorption onto heparin-containing polymer relative to that onto initial non-modified polymer [64]

Protein	Concentration of protein in solution, mg/ml	Silicone rubber	Polypropylene
Fibrinogen	3.8	1.3	1.6
γ-Globulin	10.0	0.86	1.6
β-Lipoprotein	4.0	0.70	1.4
α-Lipoprotein	4.0	4.50	0.9
α-Globulin	4.0	0.78	1.43
Albumin	20.0	0.65	1.2
Thrombin	0.1	24.0	3.8
Hageman factor	0.01	—	1.7

fibrinogen were 0.9 + 0.1 mg/cm^2 and 3.0 + 0.2 mg/cm^2, respectively. In Refs. [106] and [107] the binding of some plasma proteins to immobilized heparin was evaluated. The proteins were serum albumin and fibrinogen that were intrinsically present in plasma; thrombin and plasmin arrived via activation of the clotting and fibrinolytic systems of the blood. According to their binding strength with immobilized heparin, these proteins can be ordered in series: fibrinogen $\simeq$ thrombin $>$ plasmin $\gg$ albumin. It is noteworthy that a qualitatively similar dependence was observed for heparin in solution [116]. The complexes of immobilized heparin with fibrinogen and thrombin are rather stable (the corresponding association constants for the complexes with fibrinogen and thrombin are $(1.4 \pm 0.5) \times 10^5\ M^{-1}$ and $(8 \pm 2) \times 10^5\ M^{-1}$) and are not subjected to dissociation in physiological solution.

Thus, the increased platelet adhesion observed upon heparinization is a consequence of the enrichment of the surface of a HCP by fibrinogen [108]. Fibrinogen, as was already stated above, may specifically interact via its oligosaccharide fragments with the enzymes of the platelet membrane.

Hence, concerning the interaction with plasma proteins, covalently immobilized heparin performs identically to heparin in solution, and this results in the enrichment of the HCP surface with the most thrombogenic plasma components: fibrinogen and thrombin.

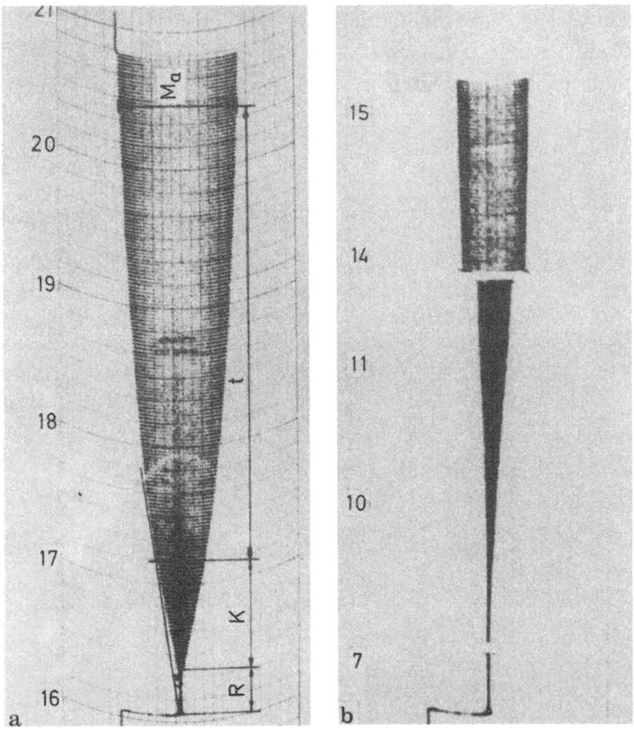

Fig. 6a and b. Thromboelastograms of canine blood before (a) and after (b) contact with heparin-containing hydrogel (4 ml of blood were incubated with 2 g of gel containing 11.8 mg of covalently immobilized heparin per g) [108]

Table 15. Parameters derived from thromboelastograms of blood before and after exposure to heparin-containing polymer [108]

Parameter		Before contact with blood	After contact with blood
R	— time corresponding to invisible phases of clotting process, thromboplastin and thrombin formation, cm	2.0	19.0
K	— time at which a fibrin clot begins to form, cm	4.7	17.5
M_a	— a characteristic of fibrinogen concentration, of the number and functional activity of platelets, and of fibrin stabilizing factor, cm	5.0	3.0
t	— a characteristic of the rate of fibrin clot formation, cm	21.7	17.0
E	— clot elasticity, $E = 100 \cdot M_a/(100 - M_a)$	5.3	3.1
R/K	— a characteristic of prothrombin consumption	0.43	1.09
C_i	— coagulation index	0.74	0.08
tgα	— a characteristic of the rate of clot formation	0.14	0.03

Alike heparin in solution, immobilized heparin endows the greatest effect on the cascade of enzymatic blood clotting reactions. The most characteristic changes taking place in the blood when it is brought into contact with a HCP were evaluated in Refs. 105, 107, 108 by thromboelastographic and coagulogram techniques. Typical thromboelastograms are given in Fig. 6, while Table 15 presents the results of their treatment. The contact of blood with HCP is seen to be accompanied by a marked decrease in blood coagulant activity and by a significant change of the clotting system

Table 16. Results of coagulogram investigations [107]

Volumetric ratio HCP/blood		1:4	1:2	1:1	1:1	2:1
Duration of contact, min		30	30	30	60	60
Plasma tolerance for heparin, s	I[a]	135	40	59	73	59
	II	353	125	495	1800	2100
Recalcification time, s	I	117	—	60	87	60
	II	170	150	160	290	180
Free heparin, s	I	3	—	4	4	4
	II	4	2	4	5	5
Thrombin time, s	I	21	—	14	10	14
	II	29	16	17	24	22
Heparin time, s	I	21	—	20	20	20
	II	26	25	30	45	42
Prothrombin index, % of initial		99.9	86.9	67.7	54.1	62.5
Fibrinogen concentration, mg%	I	425	440	365	510	365
	II	300	250	210	250	170
Activity of fibrinstabilizing factor, % of initial		80	—	57	41	54
Degree of thrombotest	I	IV	—	VI	VI	VI
	II	V	VI	IV	IV	III

[a] I — before contact with HCP, II — after contact with HCP

parameters. It involves the lengthening of the invisible phase of clotting — the increase of time at which the fibrin clot begins to form, and the decrease of fibrinogen concentration, of the number and functional activity of platelets, and of the activity of fibrin stabilizing factor, etc. Similar changes in the blood clotting system were observed when blood was contacted with N-vinylpyrrolidone-methyl methacrylate copolymers containing physically adsorbed heparin [117)], although in this case the elution of heparin, which was not sufficiently tightly bound to the polymer was possible.

A more detailed biochemical control of the clotting system was carried out by the coagulogram method (Table 16) [105, 107)]. Let us consider step by step the character of the changes of the observed parameters. The parameters to be treated first are the ones characteristic of the total coagulant activity of plasma which are the most important. These include plasma tolerance for heparin and recalcification time. The first one is markedly decreased and the second one increased after the fluid is contacted with the HCP gel, which indicates the decrease of the total coagulant activity. Parameters of the second phase, except of the concentration of heparin, are also changed quite substantially. The decrease of the prothrombin index (which is the ratio of the prothrombin time of the initial plasma to that of the plasma under examination) indicates that the activity of one or more factors of the prothrombin complex (II, V, VII and X) is reduced; an increase in heparin time indicates an increased tendency to hypocoagulation. Vast changes were observed in the third clotting phase. For instance, the thrombin time, which is the function of concentrations of fibrinogen and of the thrombin inhibitors, is slightly increased; the fibrinogen concentration is significantly decreased and the activity of fibrin-stabilizing factor is supressed. Besides this, a low degree of thrombotest was established (thrombotest involves visual evaluation of the state of fibrin: I — opalescence, II — fine fibrin "groats", III — fibrin flakes, IV — fibrin strands, V — a network of strands, occupying flakes, VI — dense fibrin strands, VII — a network of strands, occupying the volume of the fluid in the tube, VIII — a fine network sack, IX — a dense fibrin sack, occupying the whole tube).

Similar changes of the basic parameters of the blood clotting system were reported in actually all of the works dealing with the interaction of HCP with blood and its components. HCPS adsorb to the largest extent thrombin [32, 96, 118, 119)], fibrinogen [65, 94)], factors IX, X, Xa, XI [23, 94, 99, 120)] and antithrombin III [64, 94)], the last one being the only protease adsorbed from plasma. The adsorbed thrombin retains its activity and

Table 17. Results of hemoperfusion through a column containing 865 mg of heparin immobilized in polyacrylamide gel [121)]

Sample	Blood clotting time, min	Fibrinogen, mg %	Platelets, thousands	Thrombin time, s	Heparin time, s
Initial	11	332	220	15	5.5
At the outlet of the column (after perfusion)	> 30	242	180	26	18.0
From the femoral artery after the blood has passed through the living body	11	311	210	12	5.0

is inactivated only by antithrombin III. Inactivation is substantially accelerated by immobilized heparin [32].

The presented results convincingly demostrate that the blood clotting system is strongly affected by HCPs even when heparin is not eluted into the blood at all. We thus arrive at the question concerning how the arising clotting factor deficiency effects the living organism, which implies in particular whether the observed changes of the clotting system are dramatic or not for the living body as a whole? The answer to the question was given in Ref. 121 where the interaction of canine blood with HC-hydrogel was examined (Table 17). The blood was perfused through a column filled with HCP gel which was connected with its inlet part to the dog's femoral artery and the outlet part to the femoral vein. As is illustrated by Table 17, the animal's organism quite successfully overcomes the changes in blood that accompanied its contact with the polymer of sufficiently high heparin content: after the total recirculation through the animal's body, the blood clotting time was restored to the value characteristic for the operated animal before operation. The animal subjected to perfusion was controlled in a post-operative period for a month and no pathological changes having been observed.

The lack of unfavorable effects of HCP on a living body was later-on confirmed by the experiments on implantation of Dacron vascular grafts modified by covalently bound heparin [105, 108]. The grafts were implanted into dogs in the abdominal region of the aorta and in place of the fragment of the carotid artery. The observation times were from 2 weeks to 6 months. In all cases, the grafts were found to be patent and free of thrombi. After 3 months, the internal wall of the prostheses was covered with neointima having grown from anastomoses. No deviations from the normal behavior of the animals were conspicious during the whole period of observation. Non-heparinized prostheses in similar conditions were entirely clotted within 3 days of implantation.

In summary, it may be stated that the immobolized heparin interacts with some blood clotting factors, acting thus as a common anticoagulant. Modification of polymers by covalent immobilization of heparin may result in substantial and, what is even more important, chronic (long-term) thromboresistance enhancement. What is called as thromboresistance is the property of the surface of a material providing a significantly slower thrombus growth than the growth at the surface of connective tissue (neointima). A condition required for immobilized heparin to exhibit its anti-

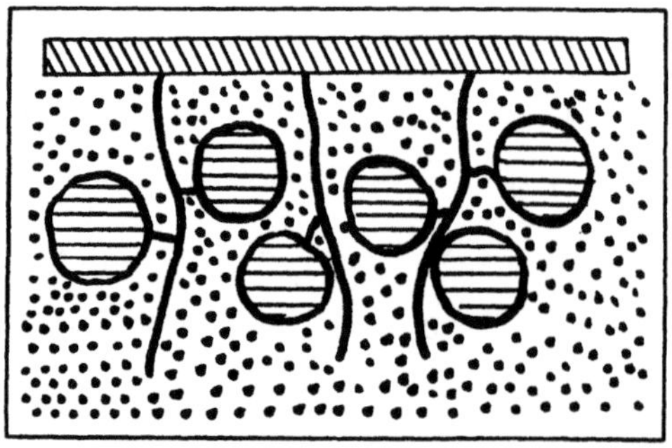

Fig. 7. Schematic presentation of the structure the of modified polymer. Solid lines — chains of hydrophylic polymer; dashed regions — macromolecules of heparin; dots — water [105]

coagulant properties is the sufficient mobility of heparin macromolecules and their accessibility for physiological substrates. This task is reached in the best way by chemical modification of the polymer surface with a hydrogel containing covalently immobilized heparin (Fig. 7) [105–108,121].

5 Mechanism of Enhanced Thromboresistance of Polymeric Materials in the Presence of Heparin

The results discussed in the previous Section permits to make some assumptions regarding the mechanism of enhanced thromboresistance of HCP. Two self-excluding viewpoints concerning the problem coexist in the literature. The first one implies the elution of heparin from a polymeric surface into the bloodstream which prevents clotting in a way common for the anticoagulant itself [64,67–73]. According to the second, the enhanced thromboresistance constitutes a property which is intrinsic to any HCP [89,99,101,102,105–108,122–125].

What are the experimental data underlying the first hypothesis? It is common knowledge today that the covalently bound heparin (if bound to the polymer rigidly and inexorably) is usually less active than ionically attached heparin which is capable of elution. In a series of works a group of Japanese investigators, who are today probably the only supporters to this concept, has disclosed the correlation between thromboresistance of the polymer and the amount of heparin released into the bloodstream [68–71]. The minimal elution rate of heparin providing sufficient thromboresistance was found to be 4×10^8 g/cm^2 · min. The decrease of the rate of elution resulted in a drastic decrease of thromboresistance.

At the same time there are some experimental facts that contradict this hypothesis. These are the following: Under sufficiently mild conditions heparin can be covalently immobilized onto a polymeric material without loosing its anticoagulant activity [101,102]. The second fact: blood that has been stored for a few hours in a HCP vessel (and remained unclotted), clotted within 5 minutes when put into a glass vessel. Thus, although heparin was not eluted, the polymer itself was thromboresistant [64,67]. The third: polymeric materials displaying regions uncoated with heparin larger than 20 microns in diameter are thrombogenic [62]. This would not have been observed if the elution of heparin were the cause for thromboresistence. Ultimately, the amount of heparin immobilized on a polymer surface (up to 500 ng/cm^2) cannot prevent the clotting of the entire volume of blood in contact with the polymer during its exploitation [62,126]. The authors of the "elution" concept try to explain these facts by assuming that the eluted heparin is concentrated within a short distance from the polymeric surface and prevents the clotting of only the boundary layer of blood. Thus, the very assumption conceals the admission of thromboresistance enhancement owing to immobilized but not the leaking heparin regardless of the forces keeping heparin at a polymer surface. Elution of heparin may probably be added to the thromboresistance enhancement, but the specific role of heparin in this case is not clear:

desorption of some compounds (surface-active compounds, heparin, albumin, etc.) from polymeric surface may also retard thrombus formation [4].

A concept of thromboresistance enhanced by *covalently* bound heparin seems to have a more sound experimental background and to be less contradictory. The majority of experimental data presented in the previous section unambiguously indicate that elution of heparin is not necessary for a high thromboresistance to be exhibited. Sufficient mobility of the heparin molecules is required as well as their accessibility for substrates. In this respect, HCPs may be contemplated as analogs of the internal wall of the blood vessel (vas intima), which naturally contains heparin and other mucopolysaccharide anticoagulants and carries a negative charge [1,127]. As was demonstrated in a vast number of studies, the proteins adsorbed onto HCP surfaces are neither denaturated, nor is the Hageman factor activated [64,65,88]. Nevertheless, it should be noted that the non-activation of the Hageman factor by itself does not guarantee adequate thromboresistance, as there are polymeric materials containing tertiary ammonium groups that do not activate this factor but are still thrombogenic [4], What probably substantially contributes to enhanced thromboresistance of HCP is that the adhered platelets are not activated [4,78], i.e., no active coagulants are released into the bloodstream. To a certain extent this effect may be attributed to the repulsion between the blood cell elements and the heparinized surface, which are both negatively charged.

In addition, immobilized heparin may react with the most important blood clotting factors, thus neutralizing their effect. Two questions may arise, however. The first one concerns the possible results of chronic *in vivo* studies of HCP which appear puzzling in sight of limited amounts of immobilized heparin suggesting its inactivation as soon as the anticoagulant gets saturated with the clotting factors. One may assume that the immobilized heparin may act as a catalyst accelerating the inhibition of one or another clotting factor. This assumption was verified in Ref. 122, which demonstrated that the affinity of antithrombin III to the immobilized heparin was decreased on binding of adsorbed antithrombin by thrombin: the antithrombin III-thrombin complex was eluted from the polymer with 0.5 *M* NaCl solution, while antithrombin itself can be eluted with 0.8 *M* NaCl solution. The reaction of immobilized anticoagulant with antithrombin III, acceleration of antithrombin III-thrombin reaction (similarly as in solution) and separation of inactive antithrombin III-thrombin complex from heparin may be considered as one of the possible pathways of the catalytical action of the immobilized heparin.

Why the fibrin clot is not formed at the surface of a HCP is the second question. This surface is highly enriched with adsorbed fibrinogen which inevitably gets into contact with thrombin formed at the surface. Two explanations of this phenomenon may be given. The first one assumes that in the presence of the immobilized heparin, the inactivation of thrombin with antithrombin III is so rapid and complete that there is actually no enzyme left to affect fibrinogen proteolysis. The alternative explanation may involve the two reactions running parallell: the inactivation of thrombin with antithrombin III and interaction of thrombin with fibrinogen. The answer to the second question, i.e., why the fibrin clot is not formed was disclosed in Refs. 105–108. It was shown that alike the complexes of the heparin in solution [128–131], the complexes of immobilized heparin with the thrombogenic proteins-thrombin and fibrinogen display similar lytic activity towards unstabilized fibrin.

Fibrin clot formation is known to proceed according to the following scheme [116]:

Fibrinopeptide

Fibrinogen → Fibrin-monomer ⇄ Fibrin-polymer ⇄ The agregates

Fibrin-stabilizing factor $\xrightarrow[Ca^{2+}]{\text{thrombin}}$ Activated fibrin-stabilizing factor

The agregates $\xrightarrow{Ca^{2+}}$ Stabilized fibrin

The first step of the process is the enzymatic hydrolysis of fibrinogen with thrombin of the arginine-glycine bonds, fibrinopeptides and fibrin-monomer molecules being formed. Succeeding steps involve polymerization of the fibrin monomer and subsequent aggregation of the polymer via interaction of the oppositely charged ionic groups distributed along fibrin-monomer molecules. Both reactions are reversible, the equilibrium constants being $(16.03 \pm 0.72) \times 10^6\ M^{-1}$ and $(3.66 \pm 0.42) \times 10^6\ M^{-1}$, respectively. Covalent crosslinking of the unstabilized fibrin involving amino groups of the lysine residues and carboxy groups of the glutamic acid residues is then effected by the activated fibrin-stabilizing factor. As was demonstrated in Refs. 105 and 107 (see also Table 16) immobilized heparin, on the one hand, suppresses the activity of the fibrin-stabilizing factor, while, on the other hand, it forms complexes with the same proteins as heparin in solution, viz. with fibrinogen and thrombin, which display the same physiological functions as complexes of heparin in solution (Fig. 8) [105-108]. In fact, Fig. 8 demonstrates that the complexes of immobilized heparin with fibrinogen, thrombin, and plasmin display lytic action on unstabilized fibrin (the amount of proteinaceous compounds in solution over unstabilized fibrin precipitate is increased). The low lytic effect of the immobilized heparin-plasmin complexes is accounted for by the low content of this enzyme in the complex (two orders lower than of fibrinogen).

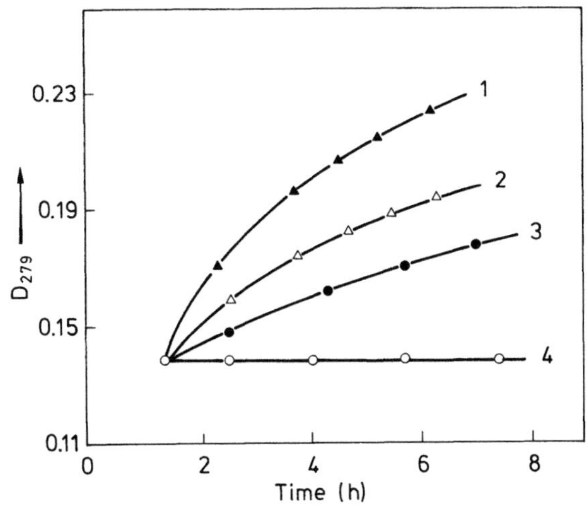

Fig. 8. Kinetics of unstabilized fibrin hydrolysis with complexes of immobilized heparin with fibrinogen (1), thrombin (2), fibrinolysin (plasmin) (3), and serum albumin (4) [106]

The kinetics of the lytic effect displayed by the complexes of immobilized heparin with thrombin and fibrinogen, in distinction from those with plasmin, are described by their saturation curves. The observed slowing down of the dissolution of unstabilized fibrin is probably due to the inhibiton of the lytic activity of the complexes by the soluble products of the reaction. In fact, as it was shown in Ref. 106, further addition of immobilized heparin-protein complex to partially hydrolyzed fibrin results in a complete recovery of the dissolution rate.

Although the mechanism of the lytic action of the complexes of immobilized as well as in fact of the soluble heparin, is in many aspects not yet clear (the complexes are assumed, for instance, to interact with fibrin-monomer causing a shift of the reaction of fibrin formation towards dissolution of the fibrin-polymer aggregates) there are some points that may be stated unequivocally. Upon reacting with the most thrombogenic plasma proteins, viz. thrombin and fibrinogen, immobilized heparin does not only neutralize their direct functions (active thrombus "builders"), but transforms them into rather efficient lytic agents. The latter, in view of the substantial supression of the fibrin-stabilizing factor activity, certainly adds to the total thromboresistance enhancement.

Thus, immobilized heparin acts as a multipurpose polymer modifier which may inhibit the blood clotting process at actually every stage and is even able to hydrolyze unstabilized fibrin. In other words, enhanced thromboresistance is an intrinsic property of HCP which can be regarded as hypocoagulation of the blood layers adjacent to the polymer. The blood in this layer is in a pseudoheparinized-like state which differs from the truly heparinized in that heparin, although present in blood and performing all of its natural functions, is tightly bound to a solid polymeric matrix. The latter results in a fast recovery of all clotting system parameters in the living body as soon as the blood is no longer in contact with the polymer. Immobilized heparin may execute its anticoagulant properties only if it is tightly bound to the polymer surface and under favorable conditions for its functioning. Such conditions are realized in polymeric hydrogel formed from hydrophilic monomers.

6 Some Approaches to Increase the Efficiency of Immobilized Heparin Action

The above-described results have demonstrated that immobilized heparin may affect many of the steps of the blood clotting process. However, such steps as platelet adhesion and lysis of unstabilized fibrin are not connected directly with heparin action. Although, as we have already stated, the adhered platelets are usually not activated, this fact cannot exclude the possibility of the formation of the "white thrombus" which is an aggregate of platelets [132–137]. This is why the search for methods reducing platelet adhesion onto HCP surfaces is now attracting much attention. One of the first approaches to solve this problem was the treatment of HCP with antithrombin III [138]. This inhibitor is strongly associated with heparin, preventing thus its alternative complexing with other proteins, namely fibrinogen and thrombin, which promote platelet adhesion onto HCP.

The second group of methods aiming at the reduction of platelet adhesion is based on the phenomenon discovered by Kim et al.[139] and Jorgensen and Stoffersen[140] who observed that modification of the polymer surface with serum albumin (albumination) decreased the number of adhered platelets. Simultaneous modification of polymers with heparin and albumin is performed by either preliminary adsorption of albumin, subsequent adsorption of heparin, and crosslinking the heparin and albumin molecules with glutaric dialdehyde[141], or adsorption on a polymer surface of a heparin-albumin conjugate[142]. Such conjugates are obtained by interlinking heparin and albumin with water-soluble carbodiimide. They contain 10–16 weight % of bound heparin, whose anticoagulant activity, as determined by measuring the thrombin time was higher by a factor of 1.25 compared to heparin in solution, whereas this activity evaluated by the ability to inhibit the factor Xa was only 50% of the activity of heparin in solution. The polymers modified in this way are not only able to inhibit the blood clotting process, but also have a much lower tendency to platelet adhesion[143].

The increased platelet adhesion onto HCP surfaces is mainly due to the high surfacial concentration of fibrinogen associated with immobilized heparin. Therefore one of the solutions to the problem could have been provided by simultaneous immobilization of heparin and proteolytic enzymes which could hydrolyze the adsorbed fibrinogen. The immobilized enzyme could also provide the lysis of the fibrin clot, in case it was formed, for instance in stagnation regions of the complicated devices.

At present, the binary water-soluble preparation of heparin and proteolytic enzymes is being applied for the treatment of thromboses. For instance, injection into the bloodstream of heparin-plasmin complex or a heparin-plasmin-streptokinase preparation leads to the total dissolution of the thrombus, while if introduced separately, heparin and streptokinase do not display the lytic action at all, and plasmin, alone or together with streptokinase, dissolves the thrombus only partially[132]. The treatment of acute thrombophlebitis with trypsin resulted in a full dissolution of the thrombus and in an increase of antithrombin III in the blood[133]. Administration of trypsin together with heparin has an effect similar in efficiency to the action of the heparin-plasmin complex[134]. The use of a mix of heparin and urokinase for improving thromboresistance of polymeric materials was also described[135]. These substances were immobilized by preliminary coating of the surface of a polymer with a graphite layer and subsequent adsorption of heparin and the enzyme.

Table 18. Thromboresistance of polyethylene modified with heparin and trypsin[137]

Concentration of immobilized substance, mg/cm²		RBCT ± 10%	RIPA ± 15%
Heparin	Trypsin		
0	0	1.2	1.8
0.3	0	1.8	3.9
1.5	0	10.0	6.0
0	0.1	1.3	1.4
0.3	0.2	2.0	2.2
0.3	0.3	3.4	1.5

Table 19. Coagulogram investigations [136]

Content of physiologically active substance in the polymer, mg/g		Heparin 1.9	Heparin 3.2 + trypsin 0.5	Heparin 1.5 + trypsin 1.6
Plasma tolerance for heparin, s	I[a]	73	73	73
	II	80	93	∞
Recalcification time, s	I	44	44	44
	II	69	91	∞
Prothrombin index, % of initial		72	90	71
Thrombin time, s	I	20	20	20
	II	25	135	∞
Activity of fibrin-stabilizing factor, % of initial		56	11	No clot formed
Fibrinogen concentration, mg %	I	250	250	250
	II	200	110	40

[a] I - before contact, II — after contact

The easiness of elution of these compounds into the bloodstream does not permit an unambiguous interpretation of the obtained results. For this very reason there is actually no hope for successful long-term implantation of polymeric materials modified in this way.

More extensive studies were carried out on polymeric materials modified with a hydrogel containing covalently immobilized heparin and trypsin [136, 137]. Table 18 compiles the results of *in vitro* tests of the thus-modified low-density polyethylene. Obviously, co-immobilization of heparin and proteolytic enzyme onto polyethylene results in an increase of the blood clotting time and decreases the number on adhered platelets as compared to immobilization of heparin alone.

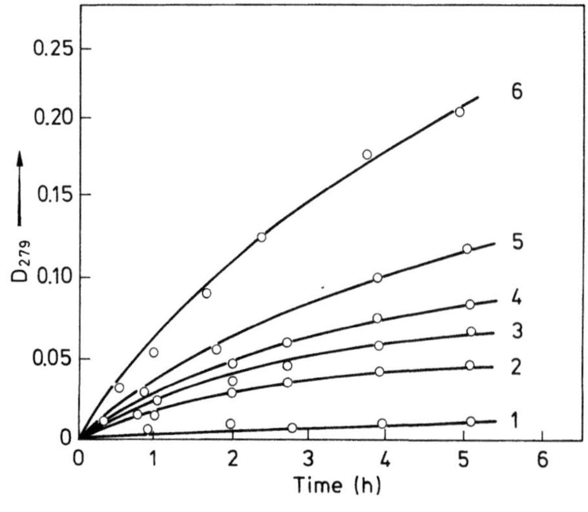

Fig. 9. Kinetics of stabilized fibrin lysis by gels containing (mg/g) 2.95 heparin (1), 0.46 trypsin (2), 1.46 heparin + 0.29 trypsin (3), 3.15 heparin + 0.46 trypsin (4), 1.29 trypsin (5), 1.29 trypsin + 1.5 heparin (6) [136]

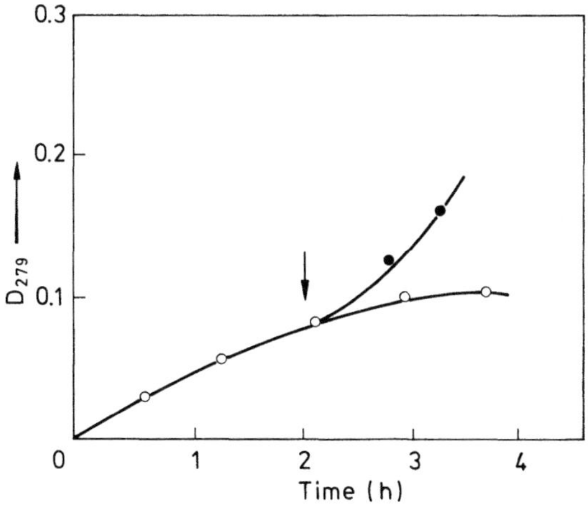

Fig. 10. Kinetics of stabilized fibrin lysis by the (heparin + trypsin)-containing polymer. The arrow marks the addition of the polymer containing immobilized heparin only [137]

Acute and chronic experiments on implantation of vascular fluorlon-poly(ethylene terephthalate) grafts (10–12 mm in diameter, 35–40 mm long) modified with heparin and trypsin confirm, in general, rather high thromboresistance of such polymers predicted in *in vivo* tests. The grafts were implanted into the internal chest section of the caudal vein of a dog where the blood flow is very slow and there is high tendency for thrombosis. The vascular grafts were entirely patent for over 7 months, while the unmodified grafts were totally clotted within 6–8 hours.

To clarify the mechanism of enhanced thromboresistance of such polymers, the state of the blood clotting system before and after the contact with the polymer was examined [136] (Table 19). Obviously, the changes of the blood clotting system parameters are of the same type as those accompanying the interaction of blood with HCP (see Table 16) but essentially exceeding the latter. Consequently, the heparin-protease-

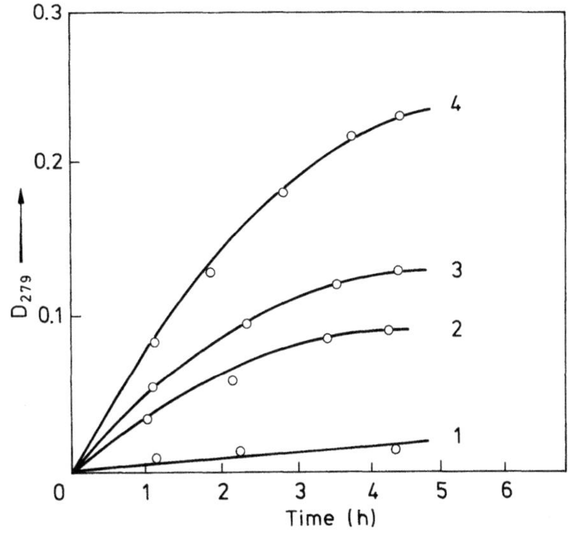

Fig. 11. Kinetics of fibrinogen hydrolysis by the polymers containing (mg/g): 1 – 2.95 heparin, 2 – 0.04 trypsin, 3 – 0.12 trypsin, 4 – 0.12 trypsin + 2.95 heparin [136]

modified polymers, being brought into contact with blood, act as active anticoagulants, and, moreover, display fibrinolytic action (Fig. 9).

Fig. 9 demonstrates the proportionality between the amount of immobilized trypsin and the lytic activity of the trypsin-containing material. The higher lytic efficiency of the binary trypsin-heparin systems seems unexpected (cf. curves 2 and 4, and 5 and 6 of Fig. 9), however. Thus, immobilized trypsin and heparin of the described polymeric materials mutually affected each other. This is manifested by the enhanced anticoagulant action of heparin promoted by trypsin and by greater lytic efficiency of trypsin provoked by heparin.

The observed enhancement of the catalytical efficiency of immobilized trypsin in the presence of immobilized heparin was shown to be due to the association of the products of fibrinolysis (as in the case of fibrinogen) which removed them out of reach of immobilized trypsin [137]. In fact, as is seen from Fig. 9, the lysis rate gradually decreases as the reaction products are accumulated, i.e., accumulation of the lysis products causes the "poisoning" of the catalyst. Addition of new polymer, but this time of that containing immobilized heparin only, again leads to an increase of the lysis rate (Fig. 10).

The more profound and extensive effect of immobilized heparin on the blood clotting system observed in the presence of immobilized trypsin is due to the tryptic lysis of the protein constituents of the complexes of immobilized heparin with the most thrombogenic plasma proteins (thrombin and fibrinogen) (Fig. 11).

Actually, the polymers modified with heparin and trypsin are (Fig. 11) more vigorously hydrolyzing the fibrinogen than the polymers containing trypsin only. This is explained by the association of fibrinogen with immobilized heparin resulting in a higher local concentration of fibrinogen within the reach of the immobilized trypsin and, there for in a rise in the rate of hydrolysis of fibrinogen.

Polymers modified with two co-immobilized physiologically active substances may

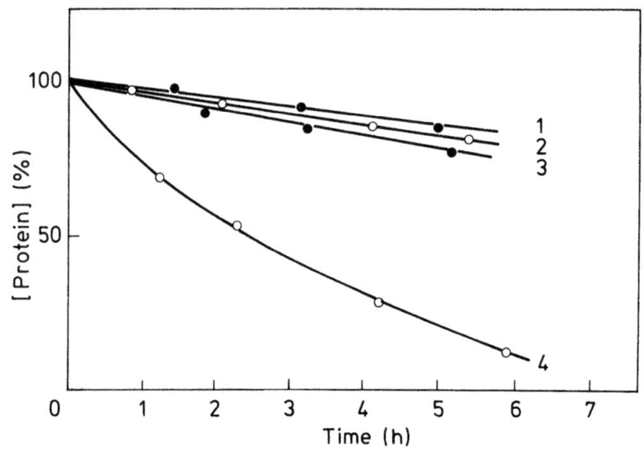

Fig. 12. Concentrations of fibrinogen (○) and serum albumin (●) versus time of incubation with the polymers containing 0.46 mg of trypsin/g of the polymer (1,2) and 0.51 mg trypsin/g polymer + 1.43 mg heparin/g polymer (3,4) [136]

thus be regarded as a macromodel of a proteolytic enzyme in which the immobilized heparin acts as the substrate-binding portion while immobilized trypsin is the active center itself. Such a combination must enhance the selectivity for fibrinogen hydrolysis when fibrinogen is mixed with the other proteins non-interacting with immobilized heparin (Fig. 12). Figure 12 shows that the rate of fibrinogen hydrolysis by the trypsin and heparin co-modified polymers, substantially exceeds the rate of hydrolysis of serum albumin by these polymers as well as the rate of hydrolysis of fibrinogen by the polymer containing immobilized trypsin only.

The hydrolysis of fibrinogen absorbed by immobilized heparin accounts for the decreased platelet adhesion onto modified polymers derived from HCP by additional immobilization of trypsin (see Table 18).

On the other hand, the nature of the substances in use (a protein and a polysaccharide) predetermines the probable side-effects which makes it impossible to predict *a priori* the results of implantation of the resultant materials, particularly when implanted constructions have a large surface area.

In fact, during the chronic contact with blood, the modifying substances must be either degraded by the enzymes or inactivated by the inhibitors which are present in blood. Although the prevention of thrombosis during the initial period of implantation (days and weeks) was shown to be of substantial importance, subsequent growth of the connective tissue isolating the implantate from blood may be such that the size of the implanted construction may play a crucial role in estimating the applicability of the modifying substance. For sufficiently large constructions, the time of encapsulation may significantly exceed the time of the active life of the modifier.

This and other side effects may in principle be overcome by following two pathways. The first involves the use of synthetic analogs of heparin. This approach was first applied in 1951, the synthetic analogs were obtained by synthesis of polystyrene sulfonic acid [144]. Upon sulfonation of polystyrene, the blood clotting time was shown to be reasonably increased (cf., 41 min instead of 29 min at the surface of the unmodified polystyrene), which gave rise to the development of this route predominantly via the synthesis of sulfonated derivatives of natural and synthetic polymers (polysaccharides, in particular). These polymers were shown to be capable of modelling some functions of heparin and even displayed improved thromboresistance properties [145–146]. Their implantation, however, was a failure, as it was accompanied by severe side-effects, e.g., blood hemolysis, whereas the activity of most of the sulfonated polymers was almost one order of magnitude lower than that of heparin. Moreover, the detailed study of the heparin-antithrombin III interaction has led [27] to the conclusion that, because of the high specificity of the said interaction, no other polysaccharide could be an adequate substitute for heparin.

In this regard, let us cast a glance at the approach that we have suggested for making protective coatings on the surface of polymeric materials [147]. The idea underlying this approach consists in that the polymer is chemically modified with a biospecific adsorbent that contains a low-molecular biospecific ligand which, consequently, is less subjected to enzymatic degradation. The ligand is immobilized within a hydrophilic polymer and it may interact with only one class of compounds. On contact of this modified polymer with a multi-component mixture of various compounds, the compound that is capable of interacting with the attached ligand will be immobilized. The difficulty of the immobilization of heparin using the described procedure, consists in the extremely low concentration of this anticoagulant in the blood. The practical application of polymeric materials with covalently immobilized biospecific sorbent for heparin must then involve periodic injection of heparin into the bloodstream. Heparin is concentrated at the polymer surface. The heparin immobilized in an agarose gel was known to absorb cholesterol, lipoproteins, and some other compounds from the blood [148]. We have suggested the reverse procedure: the sorption of heparin by cholesterol-containing polymers [147]. Two systems were examined: (1) the hydrogels produced by copolymerization of acrylamide and N,N′-methylene-bisacrylamide with the unsaturated cholesterol derivatives (UChD) of the general formula:

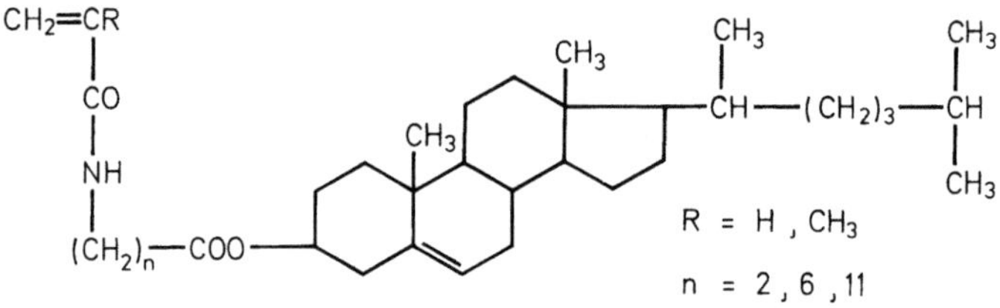

and (2) polyethylene that was modified by γ-irradiation-induced grafting of acrylamide and UChD.

Table 20 presents the results pertaining to the synthesis and properties of the gels. The gels are seen to sorb heparin from plasma solution. Their capacity (the maximal amount of heparin sorbed) increases as the mobility and accessibility of the cholesterol fragment for the macromolecules of heparin is increased by varying the length of the polymethylene spacer. The data in Table 20 illustrate that the capacity varies from 0.7 to 0.9 and 1.3 mg/mg of immobilized UChD for cholesterol esters of N-methacryloyl-β-alanine (n = 2), N-acryloyl-ω-aminoenantic acid (n = 6), and N-methacryloyl-ω-aminolauric acid (n = 11), respectively.

Table 20. The properties of cholesterol-containing hydrogels [147]

UChD and its concentration in the gel, mg/g gel + 10%	The amount of heparin sorbed, mg/g gel + 20%	Clotting time (2 ml of blood +1 g of the gel), min ±10%			
		I[a]	II	III	IV
Cholesterol ester of N-methacryloyl-β-alanine, 9.0	2.0	14	9	13	16
	6.0[b]	27	—	30	28
Cholesterol ester of N-acryloyl-ω-aminoenantic acid, 13.0	0.5	11	—	20	18
	4.0	25	—	25	27
	12.0[b]	> 30	8	> 30	> 30
Cholesterol ester of N-methacryloyl-ω-aminolauric acid, 11.0	1.0	11	—	10	11
	4.0	27	9	> 30	25
	14.0[b]	> 30	8	> 30	30

[a] I — the gel was pretreated with a 1% plasma solution of heparin, II — the sorbed heparin-containing gel washed with the saturated NaCl solution, III — the washed gel repeatedly treated with a plasma solution of heparin, IV — the gel subjected to a five round cycle "washing off — heparin sorption"

[b] The maximal amount of heparin sorbed

The sorbed heparin fully retains its anticoagulant activity: the thrombin time for both sorbed heparin and heparin in solution is 20 ± 2 s/mg heparin. The clotting time of blood in contact with the gel is proportional to the amount of heparin sorbed and significantly higher than that characteristic for the non-heparinized gel. Heparin may be repeatedly sorbed after it has been removed with saturated NaCl solution.

Similar results were obtained when polyethylene was modified by grafting acrylamide and cholesterol ester of N-methacryloyl-β-alanine to get 0.3 ± 0.1 mg UChD/cm^2. The efficiency of UChD bound to polymer was close to that of UChD immobilized in a polyacrylamide gel: the graft copolymers sorbed from the plasma solution the amount of ~ 0.9 mg heparin/mg of immobilized UChD. The clotting time of blood contacting the modified (but not heparinized) polyethylene was 6 min, whereas on its heparinization the time increased to 30 minutes. The latter figure was actually unaltered after sixfold repetition of the heparin sorption, subsequent washing heparin off the polymer and its repeated sorption.

The results presented demonstrate the feasibility of polymeric materials that are capable of repeated sorption of heparin without any loss of its activity.

7 Conclusion

In summarizing the presented experimental results, one may come to the following conclusions: There do not exist as yet any synthetic polymers known which are not exerting some type of interaction with the blood components or affecting the blood clotting process to some degree. Hardly will such polymers be obtained in the nearest future. Consequently, one of the practically sound approaches to create thromboresistant polymers may involve their surfacial modification with a compound affecting blood-polymer interaction, sharply decreasing the blood clotting activity in layers adjacent to the polymer surface. One of the few such compounds capable of performing this function and affecting actually every stage of the blood clotting process is a natural blood anticoagulant, heparin. By its immobilization (alone or in combination with other physiologically active substances), a series of polymeric materials has already been synthesized. On the other hand, there exists a number of problems, primarily related to prediction of the exact lifetimes of the modified polymers within the living body. Therefore, fundamental studies of the effects of immobilized heparin on each stage of the complicated process of blood clotting are of vital necessity.

8 References

1. Bokros, J. C.: J. Thorac. Cardiovasc. Surg. *64*, 826 (1972)
2. Vacik, M., Kliment, K., Zavadil, M.: J. Biomed. Mater. Res. *1*, 313, (1967)
3. Hoffman, A. S., Horbett, T. A., Ratner, B. D.: Ann. N.Y. Acad. Sci. *283*, 372 (1977)
4. Leininger, R. I.: CRC Crit. Rev. Bioeng. *1*, 333 (1972)
5. Kusserow, B. K., Larrow, R. W.: Trans. Amer. Soc. Artif. Int. Organs *19*, 8 (1973)
6. Platé, N. A., In: Polymeric Drugs, N.Y./London: Acad. Press 1970, p. 63
7. Platé, N. A., Valuev, L. I., Al-Nuri, M. A., Egorov, N. S.: Pricl. Biochim. i Microbiol. *13*, 673 (1977) (in Russian)

8. Platé, N. A., Valuev, L. I., Baranova, N. A., Al-Nuri, M. A., Egorov, N. S.: Pat. USSR No. 545648, Bul. Izobret. No. 5 (1977) (in Russian)
9. Platé, N. A., Valuev, L. I., Chupov, V. V.: Vysokomol. Soed. *22A*, 1963 (1980) (in Russian)
10. Marconi, W., Bartoli, F., Mantovani, E., Pittalis, F., Settembri, L., Cordova, C., Musca, A., Alessandri, C.: Trans. Amer. Soc. Artif. Int. Organs *25*, 10 (1979)
11. Gott, V. L., Koepke, D. E., Daggett, R. L., Zarnstorff, W. L., Young, W. P.: Surgery *50*, 382 (1961)
12. Panov, V. P., Ovsepyan, A. M.: Chim. Pharm. Zh. *13*, 72 (1979) (in Russian)
13. Shen, L. L., Barlow, G. H., Holleman, W. H.: Thromb. Res. *13*, 671 (1978)
14. Danishefsky, J., Steiner, H., Bella, A., Triedlanger, A.: J. Biol. Chem. *244*, 1741 (1969)
15. Cifonelli, Y. A., King, Y. I.: Carbohyd. Res. *12*, 391 (1970)
16. Ulyanov, A. M., Lyapina, L. A.: Uspechi sovrem. Biol. *83*, 69 (1977) (in Russian)
17. Fracassini, A., White, C. Y., Hunter, I. C.: FEBS Lett. *32*, 116 (1973)
18. Horner, A. A.: J. Biol. Chem. *246*, 231 (1971)
19. Yaun, L., Stivala, S. S.: Adv. Exp. Med. Biol. *52*, 39 (1975)
20. Cifonelli, Y. A.: Adv. Exp. Med. Biol. *52*, 95 (1975)
21. Ebert, C. D., Kim, S. W.: Thromb. Res. *26*, 43 (1982)
22. Zubairov, D. M.: Biochimiya svertyvaniya krovi, Moscow, Medicina 1978, p. 152
23. Gently, P. W., Elexander, B.: Biochem. Biophys. Res. Commun. *50*, 500 (1973)
24. Kurachi, K., Fujikama, K., Schmer, G., Davie, E. W.: Biochemistry *15*, 373 (1976)
25. Palos, L. A.: Haematologia *8*, 455 (1974)
26. Aranyi, P., Batke, Y., Machovich, R.: Arch. Biochem. Biophys. *181*, 678 (1977)
27. Yamaguchi, S., Hayashi, H., Hamada, F., Nakajima, A.: Biopolymers *23*, 995 (1984)
28. Björk, I., Lindahl, U.: Molecul. Cel. Biochem. *48*, 161 (1982)
29. Thunberg, L., Bäckström, G., Lindahl, U.: Carbohyd. Res. *100*, 393 (1982)
30. Stone, A. L., Beeler, D., Oosta, G., Rosenberg, R. D.: Proc. Nat. Acad. Sci. USA *79*, 7190 (1982)
31. Smith, G. F.: Biochem. Biophys. Res. Commun. *77*, 111 (1977)
32. Hatton, M. W., Regoeszi, E.: Thromb. Res. *10*, 645 (1977)
33. Kudryachov, B. A., Kalishevskaya, T. M., Lyapina, L. A.: Biochimiya *38*, 1281 (1973) (in Russian)
34. Kudryashov, B. A., Lyapina, L. A., Ulyanov, A. M.: Biochimiya *42*, 451 (1977) (in Russian)
35. Kudryashov, B. A., Ulyanov, A. M., Zhitnikova, E. S.: Dokl. Akad. Nauk SSSR *237*, 965 (1977) (in Russian)
36. Choay, J., Petitou, M., Lormeau, J. C., Sinäy, P., Casu, B., Gatti, G.: Biochem. Biophys. Res. Commun. *116*, 492 (1983)
37. McDuffe, N. M., Dietrich, C. P., Nader, H. B.: Biopolymers *14*, 1473 (1975)
38. Rosenberg, R. D., Jordon, R. E., Favreau, L. V., Lam, L. H.: Biochem. Biophys. Res. Commun. *86*, 1319 (1979)
39. Gott, V. L., Whiffen, J. D., Dutton, R. S.: Science *142*, 1297 (1963)
40. Sawyer, P. N., Pate, J. M.: Amer. J. Physiol. *175*, 103 (1953)
41. Sawyer, P. N., Suckling, E. E., Weselowski, S. A.: Amer. J. Physiol. *198*, 1006 (1960)
42. Whiffen, J. D., Gott, V. L.: J. Surg. Res. *5*, 51 (1965)
43. Whiffen, J. D.: Trans. Amer. Soc. Artif. Int. Organs *12*, 164 (1966)
44. Regelson, W.: J. Polym. Sci. *66*, 483 (1979)
45. Whiffen, J. D., Beeckler, D. C.: J. Thorac. Cardiovasc. Surg. *52*, 121 (1966)
46. Usdin, V. R., Fourt, L.: J. Biomed. Mater. Res. *3*, 69 (1969)
47. Bokros, J. C., Gott, V. L., La Grande, L. D., Fadall, A. M., Vos, K. D., Ramos, M. D.: J. Biomed. Mater. Res. *3*, 497 (1969)
48. Fourt, L., Schwartz, A. M., Becker, R. H., Kairinen, E. O.: Interac. Liquids at Solid Subst. *87*, 186 (1968)
49. Gott, V. L., Daggett, R. L., Whiffen, J. D., Koepke, D. E., Rowe, G. C., Young, W. P.: J. Thorac. Cardiovasc. Surg. *48*, 713 (1964)
50. Valiatham, M. S., Weldon, C. S., Bender, H. W., Topaz, S. R., Gott, V. L.: J. Surg. Res. *8*, 197 (1968)
51. Grode, G. A., Falb, R. D., Crowley, J. P.: J. Biomed. Mater. Res., Symp. *3*, 77 (1972)
52. Hersh, L. S., Weetall, H. H., Brown, J. W.: J. Biomed. Mater. Res., Symp. *1*, 99 (1971)

53. Leininger, R. I., Crowley, J. P., Falb, R. D., Grode, G. A.: Trans. Amer. Soc. Artif. Int. Organs *18*, 312 (1972)
54. Najiar, F. B., Gott, V. L.: Surgery *68*, 1053 (1970)
55. Bruck, S. D.: Biomat., Med. Dev., Artif. Organs *1*, 79 (1973)
56. Rea, W. J., Whitley, D., Eberle, J. W.: Trans. Amer. Soc. Artif. Int. Organs *18*, 316 (1972)
57. Eriksson, J. C., Gillberg, G., Lagergren, H.: J. Biomed. Mater. Res. *1*, 301 (1967)
58. Lagergren, H., Johansson, L., Eriksson, J. C.: Thoraxchirurgie *12*, 172 (1964)
59. Uy, S., Kammermeyer, K.: J. Biomed. Mater. Res. *3*, 587 (1969)
60. Mobin-Uddin, K., Trinkle, K. J., Bryant, L. R.: Surgery *70*, 914 (1971)
61. Mobin-Uddin, K., Utley, J. R., Bryant, L. R., Dillon, M., Weiss, D. L.: Ann. Thorac. Surg. *17*, 351 (1974)
62. Stewart, G. P., Wilkov, M. A.: J. Biomed. Mater. Res. 10, 413 (1976)
63. Leininger, R. I., Cooper, C. W., Epstein, M. M., Falb, R. D., Grode, G. A.: Science *152*, 1625 (1966)
64. Falb, R. D., Leininger, R. I., Grode, G. A., Crowley, J.: Edv. Exp. Med. Biol. *52*, 365 (1975)
65. Falb, R. D., Grode, G. A., Takahachi, M. T., Leininger, R. I.: Interac. Liquid at Solid Substr. *87*, 181 (1968)
66. Hollahan, J. R., Stafford, B. B., Falb, R. D., Payne, S. T.: J. Appl. Polym. Sci. *13*, 807 (1969)
67. Linton, A. C., Gilchrist, T., Courtney, J., Edwards, R. O.: J. Lab. Clin. Med. *89*, 724 (1977)
68. Tanzawa, H., Mori, Y., Horumiya, M., Miyama, H., Hori, M., Ohshima, N., Idezuki, Y.: Trans. Amer. Soc. Artif. Int. Organs *19*, 188 (1973)
69. Idezuki, Y., Watanabe, H., Hagiwara, M., Kanasugi, H., Mori, Y., Nagaoka, S., Hagio, M., Yamamoto, K., Tanzawa, H.: Trans. Amer. Soc. Artif. Int. Organs *21*, 436 (1975)
70. Miyama, H., Harumiya, N., Mori, Y., Tanzawa, H.: J. Biomed. Mater. Res. *11*, 251 (1977)
71. Mori, Y., Nagaoka, S., Masubuchi, Y., Itoga, M., Tanzawa, H., Kikuchi, T., Yamada, Y., Yonaha, T., Watanabe, H., Idezuki, Y.: Trans. Amer. Soc. Artif. Int. Organs *24*, 736 (1978)
72. Mori, Y., Nagaoka, S., Itoga, M., Tanzawa, H., Yamada, Y., Watanabe, H., Idezuki, Y.: Artif. Organs 2 (Suppl.), 66 (1978)
73. Mori, Y., Nagaoka, S., Tanzawa, H., Kikuchi, T., Yamada, Y., Hagawara, M., Idezuki, Y.: J. Biomed. Mater. Res. *16*, 17 (1982)
74. Mori, Y., Nagaoka, S., Kikuchi, T., Tanzawa, H.: J. Biomed. Mater. Res. *16*, 209 (1982)
75. Mori, Y., Nagaoka, S., Kikuchi, T., Tanzawa, H.: Artif. Organs *9*, 228 (1980) (in Japanese)
76. Merker, R. L., Eliash, L. J., Mayhew, S. H., Wang, J. Y. C.: Artificial Heart Program Conference Proceedings. Ed. Hegyeli, R. J., Washington: National Heart Institute Artificial Heart Program 1969 p. 29
77. Herch, L. S., Weetall, H. H., Brown, I. W.: J. Biomed. Mater. Res. *3*, 471 (1969)
78. Bruck, S. D., Rabin, S., Ferguson, R. J.: Biomat., Med. Dev., Artif. Organs *1*, 191 (1973)
79. Rittenhouse, E. A., Mohri, H., Reichenbach, D. D., Merendino, K. A.: Arch. Surg. *105*, 752 (1972)
80. Merrill, E. W., Salzman, E. W., Lipps, B. J., Gilliland, E. R., Austen, W. G., Joison, J.: Trans. Amer. Soc. Artif. Int. Organs *12*, 139 (1966)
81. Halpern, B. D., Shibakawa, R.: Interac. Liquid at Solid Substr. *87*, 197 (1968)
82. Palma, M., Riva, F., Provensale, C.: J. Polym. Sci. *15*, 2151 (1977)
83. Ferruti, P., Casini, G., Tempesti, F., Barbucci, R., Mastacchi, R., Sarret, M.: Biomaterials *5*, 234 (1984)
84. Ferruti, P., Barbucci, R., Danzo, N., Torrisi, A., Puglisi, O., Pignataro, S., Spartano, P.: Biomaterials *3*, 33 (1982)
85. Mangano, D. T.: The New England J. of Med. 894 (1982)
86. Pusineri, C., Eloy, R., Baguet, J., Paul, J., Belleville, J., Leconte, P., Farges, J. P.: Artif. Organs 5 (Suppl.), 512 (1981)
87. Hecker, J. F.: Pediatrics *4*, 467 (1981)
88. Merrill, E. W., Salzman, E. W., Wong, P. S. L., Ashford, T. P., Brown, A. D., Austen, W. G.: J. Appl. Physiol. *29*, 723 (1970)
89. Cholakis, C. H., Sefton, M. V.: ACS Polymer Preprints *24*, 64 (1983)
90. Sefton, M. V., Merrill, E. W.: J. Biomed. Mater. Res. *10*, 33 (1976)
91. Novikova, S. P.: Abstracts of 4 Vsecoyuznogo Symposiuma po Medicinskim Polymeram, Dzerzhinsk, SSSR, p. 42 (1979)

92. Labarre, D., Boffa, M. C., Jozefowicz, M.: J. Polym. Sci., Symp. *47*, 131 (1974)
93. Labarre, D., Boffa, M. C., Jozefowicz, M.: J. Biomed. Mater. Res. *11*, 283 (1977)
94. Labarre, D., Boffa, M. C., Jozefowicz, M.: Thromb. and Haemost. *41*, 346 (1979)
95. Baquey, Ch., Beziade, A., Ducassou, D., Blanquet, P.: Innov. Tech. Biol. Med. *2*(4), 379 (1981)
96. Schmer, G.: Trans. Amer. Soc. Artif. Int. Organs *18*, 321 (1972)
97. Danishefsky, J., Tzeng, F., Ahrens, M., Klein, S.: Thromb. Res. *8*, 131 (1976)
98. Hoffman, A. S., Schmer, G., Harris, C., Kraft, W. G.: Trans. Amer. Soc. Artif. Int. Organs *18*, 10 (1972)
99. Heyman, P. W., Cho, C. S., McRea, J. C., Olsen, D. B., Kim, S. W.: J. Biomed. Mater. Res. (in press)
100. Miura, Y., Aoyagi, S., Kusada, Y., Miyamoto, K.: J. Biomed. Mater. Res. *14*, 619 (1980)
101. Platé, N. A., Valuev, L. I., Gumirova, F. Kh.: Dokl. Akad. Nauk SSSR *244*, 1505 (1979) (in Russian)
102. Platé, N. A., Valuev, L. I., Gumirova, F. Kh.: Abstracts of 26 International Symposium on Macromolecules, Mainz, 1979, Vol. 3, p. 1522
103. Mester, L., Amit-Amaya, A., Mester, M.: Amer. Chem. Soc., Symp. Ser. *77*, 113 (1978)
104. Platé, N. A., Valuev, L. I.: Bullet. Exper. Biol. i Medic. *12*, 686 (1981) (in Russian)
105. Platé, N. A., Valuev, L. I., Gumirova, F. Kh., Zimin, N. K., Rozenfeld, M. A.: Voprosy Med. Chimii *28*, 97 (1982) (in Russian)
106. Platé, N. A., Valuev, L. I., Gumirova, F. Kh., Maklakova, I. A.: Voprosy Med. Chimii *28*, 19 (1982) (in Russian)
107. Platé, N. A., Valuev, L. I.: Thromb. Res. *27*, 131 (1982)
108. Platé, N. A., Valuev, L. I.: Biomaterials *4*, 14 (1983)
109. Larsson, R., Rosenberg, A., Per Olsson, P.: Thromb. Res. *11*, 517 (1977)
110. Essien, E. M., Cazenava, J.-P., Moore, S., Mustard, J. E.: Thromb. Res. *13*, 69 (1978)
111. Salzman, E. W., Merrill, E. W., Binder, A., Wole, C. F. W., Ashford, T. P., Austen, W. G.: J. Biomed. Mater. Res. *3*, 69 (1969)
112. Goosen, M. F., Sefton, M. V.: J. Biomed. Mater. Res. *13*, 347 (1979)
113. Heyman, P. W., Kim, S. W.: Makromol. Chemie (in press)
114. Noishiki, Y.: Proc. 2-nd Jap. Biomater. Soc. *65*, 112 (1980) (in Japanese)
115. Ehrlich, J.: Polymer. Eng. and Sci. *15*, 281 (1975)
116. Rozenfeld, M. A.: Dokl. Akad. Nauk SSSR *239*, 207 (1978) (in Russian)
117. Chepurov, A. K., Belomestnaya, T. M., Donezcy, Yu. A., Cidvinceva, G. A.: Bull. Exper. Biol. i Medic. *5*, 296 (1980) (in Russian)
118. Nordenman, B., Björk, J.: Thromb. Res. *11*, 799 (1977)
119. Nordenman, B., Björk, J.: Thromb. Res. *12*, 755 (1978)
120. Gitel, S. N.: Advans. Exp. Med. Biol. *52*, 243 (1975)
121. Platé, N. A., Valuev, L. I.: Abstracts of 5 Vsesoyuznogo Symposiuma po medicinskim polymeram, Riga, USSR, 1981, p. 127 (in Russian)
122. Carlström, A.-S., Lieden, K., Björk, J.: Thromb. Res. *11*, 785 (1977)
123. Goosen, M. F. A., Sefton, M. V.: J. Biomed. Mater. Res. *17*, 359 (1983)
124. Sun, A. M., Parisius, W., Macmorine, H. G., Sefton, M. V., Stone, R.: Artif. Organs *4*, 275 (1980)
125. Goosen, M. F. A., Sefton, M. V., Hatton, M. W. C.: Thromb. Res. *20*, 543 (1981)
126. Lindsay, R. M., Rourke, J. T. B., Reid, B. D., Linton, A. L., Gilchrist, T., Courtney, Y., Edwards, R. O.: J. Lab. Clin. Med. *89*, 724 (1977)
127. Gott, V. L., Whiffen, J. D., Koepke, D. K., Daggett, R. L., Boake, W. C., Young, W. P.: Trans. Amer. Soc. Artif. Int. Organs *10*, 213 (1964)
128. Kudryashov, B. A., Kalishevskaya, T. M., Polyakova, L. A.: Voprosy Med. Chimii *12*, 114 (1966) (in Russian)
129. Kudryashov, B. A., Kalishevskaya, T. M., Lyapina, L. A.: Voprosy Med. Chimii *14*, 277 (1968) (in Russian)
130. Ulyanov, A. M., Kudryashov, B. A., Lyapina, L. A.: Biochimiya *38*, 1281 (1973) (in Russian)
131. Kudryashov, B. A., Lyapina, L. A.: Voprosy Med. Chimii *21*, 165 (1975) (in Russian)
132. Shumakov, V. I., Chepurov, A. K.: Cardiologiya, *2*, 88 (1972) (in Russian)
133. Grigoryan, A. V., Gostincev, V. K., Panshin, N. N.: Chirurgiya *2*, 65 (1971) (in Russian)

134. Struchkov, V. I., Gostincev, V. K.: Aktualnye Problemy Gemostaziologii, Moscow: Nauka 1981, p. 371 (in Russian)
135. Kusserow, B. K., Larrow, R., Nickols, J.: Trans. Amer. Soc. Artif. Int. Organs *17*, 1 (1971)
136. Valuev, L. I., Chupov, V. V., Burdygina, I. F.: Abstracts of 4 Vsesoyuznogo Symposiuma po Inzhenernoj Enzim., Kiev, USSR, 1983, p. 88 (in Russian)
137. Platé, N. A., Valuev, L. I., Chupov, V. V.: Pure Appl. Chem. *56*, 1351 (1984)
138. Olsson, P., Lagergren, H.. Larsson, R., Radegran, K.: Thromb. Haemostas. *37*, 274 (1977)
139. Kim, S. W., Lee, R. D., Coleman, D., Oster, H., Andrade, J. D., Olsen, D. B.: Trans. Amer. Soc. Artif. Int. Organs *20*, 449 (1974)
140. Jorgensen, K. A., Stoffersen, E.: Thromb. Res. *17*, 13 (1980)
141. Novikova, S. P., Dobrova, N. B., Popov, S. A., Selezneva, N.: Patent SSSR No 1097339A, Bull. Izobret. No 22 (1984)
142. Hennink, W. E., Feijen, J., Ebert, C. D., Kim, S. W.: Thromb. Res. *29*, 1 (1983)
143. Hennink, W. E., Ebert, C. D., Kim, S. W., Breemhaar, W., Bantjes, A., Feijen, J.: Biomaterials *5*, 264 (1984)
144. Lovelock, J. E., Porterfield, J. S.: Nature (london) *167*, 39 (1951)
145. Fougnot, C., Jozefowicz, M., Samama, M., Bara, L.: Ann. Biomed. Eng. *7*, 441 (1979)
146. Fougnot, C., Jozefowicz, M., Rosenberg, R. D.: Biomaterials *5*, 94 (1984)
147. Valuev, L. I., Freidzon, Ya. S., Shibaev, V. P., Gumirova, F. Kh., Platé, N. A.: Dokl. Akad. Nauk SSSR *265*, 366 (1982) (in Russian)
148. Moorjani, S., Lupien, P.-J., Awad, J.: Clin. Chim. Acta *77*, 21 (1977)

Editor: K. Dušek
Received November 18, 1985

Teichoic Acids and Their Models: Membrane Biopolymers with Polyphosphate Backbones. Synthesis, Structure and Properties

Paweł Kłosiński and Stanisław Penczek
Polish Academy of Sciences, Center of Molecular and Macromolecular Studies, 90-362 Łódź, Boczna 5, Poland

Advances in Polymer Science 79

1 Introduction

Teichoic acids (TAs), naturally occurring polyphosphates of substituted glycerol (*1*) or ribitol (*2*) in the simpler cases, are biopolymers isolated from the cell walls (wall TAs) or membranes [membrane TAs or lipoteichoic acids (LTAs)] of various Gram-positive bacteria:

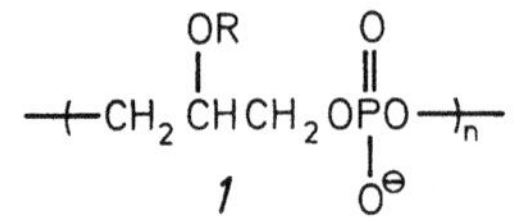

R = H, D-alanyl or sugar

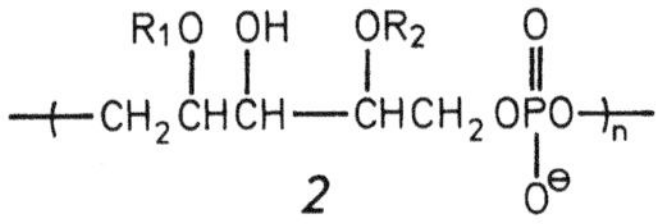

R_1 = H or D-alanyl
R_2 = H or sugar

Chemical syntheses of structures of *1* and *2*, based on the polyaddition method, were elaborated by Michelson [1] and Baddiley et al. [2], respectively.

More recently, methods of polymerization of cyclic monomers containing phosphorus became available [3], leading directly to structures like *1*.

Physiological functions of TAs in bacterial organisms [4,5], their immunological properties [6,7], and relation to some diseases (e.g. rheumatic fever) [7] are being intensively studied. Unfortunately, TAs and LTAs have mostly been isolated from bacterial sources and could not be standardized sufficiently. Therefore, structures of the isolated polymers differ considerably and are not well defined. Thus, synthetic polymers with better known structures and polymerization degrees should be welcome to this field.

The presence of hydroxyl groups in the polymer units allows to use this reactive site for covalent binding of compounds of interest, for instance biologically active compounds, whereas the presence of ionizable phosphate groups makes these polymers soluble in water.

Purified natural TAs are not immunogenic alone [7], probably due to their low molecular weights ($\overline{DP}_n \approx 40$). Unfortunately nothing is known about the metabolism of TAs in animal and human organisms and their toxicity is not clearly established [7].

In the present review, we briefly describe the structures, functions and biosynthesis of teichoic acids. Then the chemical syntheses of TAs and fragments of the polymer chains are described in more detail.

2 Structure of Teichoic Acids (TAs)

Typical wall teichoic acids are the linear polyphosphates of glycerol-1,3 or D-ribitol-1,5. Their average polymerization degree varies from 30 to 40. Carbon atoms at positions 2 and 4 in the ribitol and at position 2 in the glycerol moieties are substituted with sugar residues or with D-alanyl ester groups as shown *1* and *2* [4,8].

Sugar substituents in *1* and *2* are D-glucose, D-galactose, D-glucosamine, D-galactosamine and their N-acetyl derivatives [8,9]. The structures and extent of substitution both depend on the microorganism involved and on the conditions of bacterial growth. For example: TA from *Bacillus subtilis* consists of two kinds of chains: completely substituted with glucose and unsubstituted [10]. *Staphylococcus aureus* strain Copenhagen and *Staphylococcus albus* strain Prengel produce two kinds of polyribitol phosphate chains. One is substituted with β-D-N-acetylglucosamine, the second one has exclusively α-glycosidic bonds [11].

Some strains produce glycerol TAs with 2,3-diesterlinkages, e.g.: the teichoic acid from *Bacillus subtilis* var. *niger* WM [12] shown below:

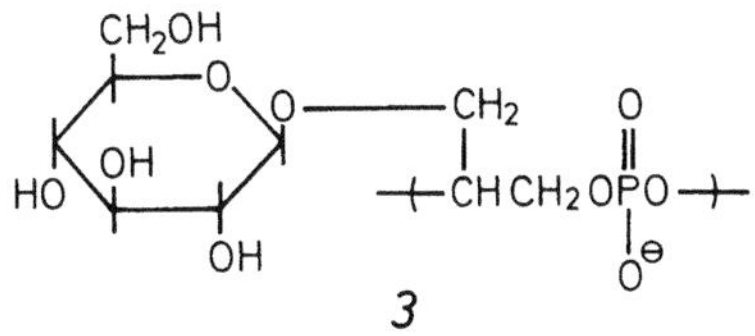

3

Acetic ester groups [13] and in one case L-lysyl residues [14] were found in TAs from Streptomyces walls as substituents instead of D-alanyl, whereas TA from *Bifidobacterium bifidum* [15] contains mannitol as the polyol constituent.

Polyribitol and polyglycerol phosphates, with sugar moieties in the main polymer chain, constitute another group of wall TAs, so-called atypical TAs (aTAs). They were isolated first from the cell walls of *Staphylococcus lactis* I3 [16]. The chains of this polymer can be represented as an alternate copolymer of glycerol and D-N-acetylglucosaminyl phosphates, as it is shown in *4*. Another aTA was extracted from *Bacillus stearothermophilus* B65. The polymer repeating unit is 1-O-α-D-glucopyranosylglycerol-3-phosphate. These units are joined by phosphoester bonds at the 6′ positions in glucose [17] (*5*):

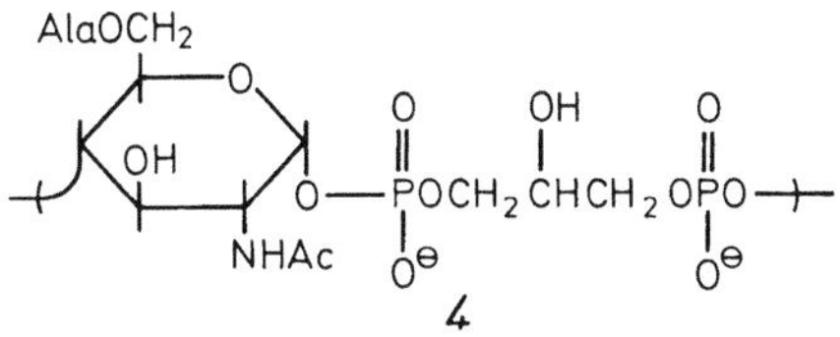

4

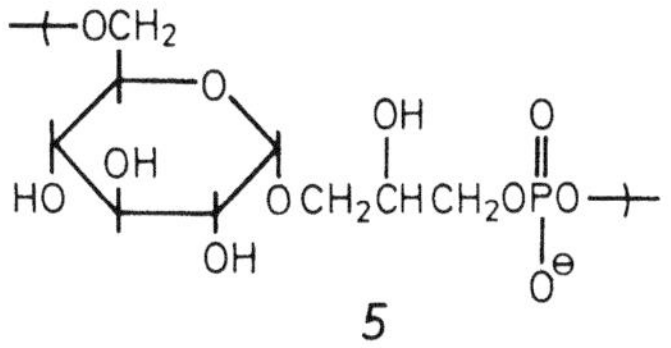

5

An atypical ribitol TA was found in a capsule of the Gram-negative bacteria *Haemophilus influenza* type a [18] (*6*). The monomeric unit of this polymer consists of 4-O-β-D-glucopyranosyl-D-ribitol-5-phosphate linked at the 4′ position in the sugar residue.

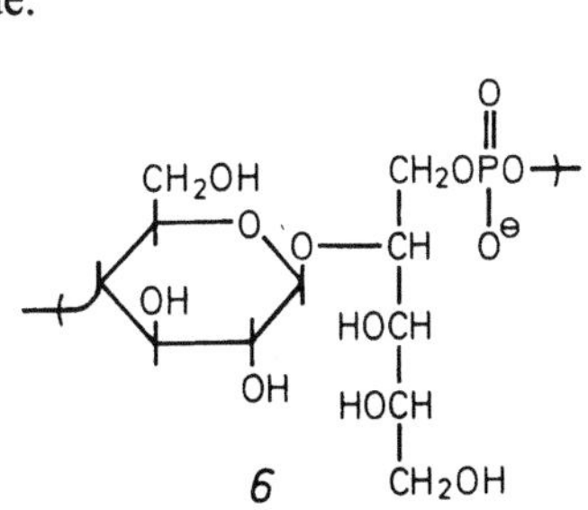

6

More complicated TAs were found in cells of *Diplococcus pneumoniae* (Pneumococcus) [15,16,19,20] and in different types of *Haemophilus influenza* [21].

Polysaccharide-1-phosphates were extracted in some cases from bacterial cell walls. For example, TA isolated from *Staphylococcus lactis* 2102 consists of ca. 23 monomeric fragments each built of α-D-N-acetylglucosamine-1-phosphate [22]. These units are linked by the phosphoester bond at the 6 position in the sugar residue, as represented by *7*.

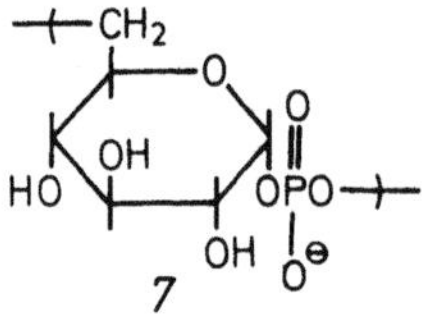

7

Lipoteichoic acids (LTAs), polymers extracted from bacterial membranes, consist of hydrophilic polyphosphate chains linked covalently with the hydrophobic lipid moieties. The LTA structures are more uniform than those of wall TAs. The LTA chains are built from polyglycerol phosphate with 1,3-phosphodiester linkages, substituted with D-alanyl or sugar residues (*1*). However, ribitol containing LTA was recently isolated from *Staphylococcus aureus* H species [23]. $\overline{DP}_n$ of these polymers equals from 28 to 35 for *Streptococcus faecalis* NCIB 8191 [24] and from 32 to 40 for *Bacillus licheniformis* [25]. The lipid residues are not structurally uniform and consist of glycolipids [26,27] and phosphatidylglycerols [25]. The lipid glycerol moieties are acylated with long-chain fatty acid esters.

3 Physiological Functions

3.1 Binding of Mg^{2+} Ions

The main physiological function of both TAs and LTAs is to collect the divalent cations, particularly magnesium ions, from the surrounding medium, and to flux them through the cell wall to the membrane region where their concentration is kept at the proper level, i.e. from 10 to 15 m*M*, by the polyphosphate chains of LTAs. The proper concentration of Mg^{2+} ions is required for the activity of enzymes associated with the membrane.

Magnesium ions are associated with the cell walls in two different ways [28]. The stronger bound cations are associated with two adjacent phosphate groups in TA chains as it is shown in *8*. The flexibility of chains and free rotation around the ester bonds allow masking the negatively charged phosphate groups by protonated amino groups of D-alanyl substituents, as it is shown in *9*. Thus, some magnesium cations could be associated with only one phosphate group and therefore their association with TA molecules is weaker. The second positive charge of Mg^{2+} in this structure is neutralized by the anion coming from the medium, e.g. Cl^-. Blocking of some phosphate groups by D-alanyl residues leads to the observed [29] deviation from the simple stoichiometric ratio P:Mg = 2.

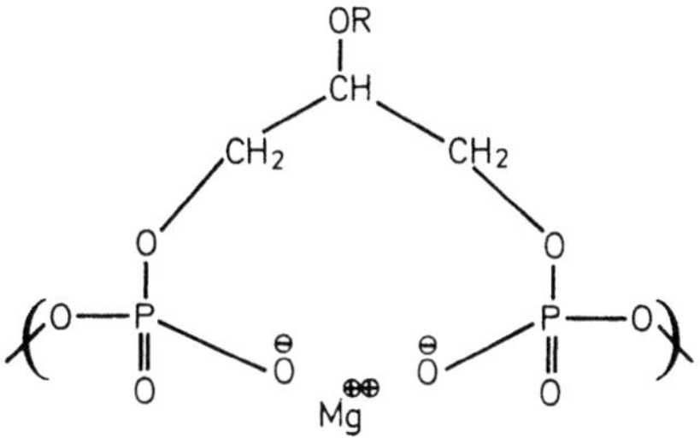

R = H, sugar
8

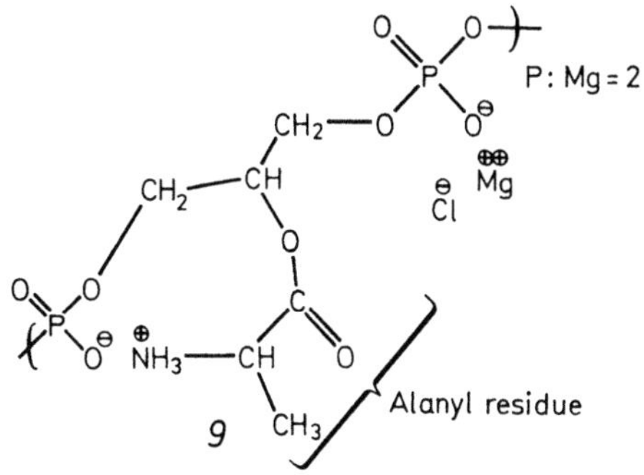

9

Glycerol TA associates with magnesium ions in solution stronger than ribitol TA. Apparent associate constants (K_a) at pH = 5 are 2.7×10^3 mol^{-1} and 0.61×10^3 mol^{-1} for glycerol [29] and ribitol [30] TA, respectively. Ribitol TA has no possibility to bind Mg^{2+} ions by two adjacent phosphate groups because of steric restrictions [30]. Ribitol TA in cell walls associates with magnesium ions with a strength similar to that observed for glycerol TA ($K_a = 2.7 \times 10^3$ mol^{-1}) [29]. Thus, magnesium ions probably form bridges between phosphate groups belonging to two neighboring TA chains. The pH of solutions has no influence on K_a but the amount of Mg^{2+} ions bound to TA molecules decreases at lower pH because of protonation of the phosphate groups. TAs have been shown to be highly selective for Mg^{2+} ion binding [29]. They bind magnesium ions even in the presence of an excess of mono- and other divalent cations.

3.2 Influence on Enzymes Activity

The teichoic acids of the wall and membrane act as a buffer system maintaining a concentration of Mg^{2+} in the range of 10–15 m*M*, which is the optimum value for the activity of enzymes associated with membranes [31]. Changes of the Mg^{2+} concentration in the medium have no effect on the activity of membrane-bound enzymes if the system keeps both the membrane and the cell wall closely in contact. Preparations devoid of cell walls (lysozyme digestion) exhibit a dependence of enzyme activity on the Mg^{2+} concentration. The membrane fragments without LTAs show a marked influence of the Mg^{2+} concentration on enzyme activity [31].

Another physiological role of both types of teichoic acids is their interaction with autolytic enzymes and the regulation of their activity [5, 32].

4 Biosynthesis

Glaser and Burger, using membrane preparations of several bacterial species and various nucleotide precursors, prepared *in vitro* a few kinds of teichoic acids which were identical with samples extracted from the cell walls of the studied strains. The following TA chain precursors were used: cytidine diphosphate glycerol (CDP-glycerol) (*10*) [33] and cytidine diphosphate ribitol (CDP-ribitol) (*11*) [34] in the case of glycerol and ribitol TAs, respectively. Glycosidation can be accomplished *in vitro* by using suitable precursors: uridine diphosphate glucose (UDP-glucose) (*12*) and nonsubstituted TA chains as substrates in the presence of active membrane preparations [35]. More effective is the glycosidation of the growing teichoic acid chains [36]. It requires the presence of the precursors of both chain and substituent in solution containing the undamaged fragments of membranes.

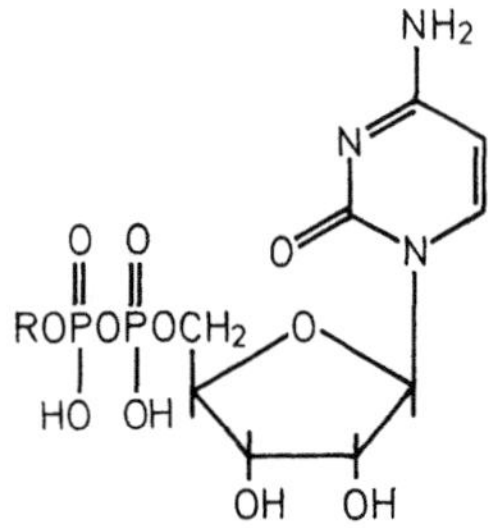

10 R = glycerol
11 R = ribitol

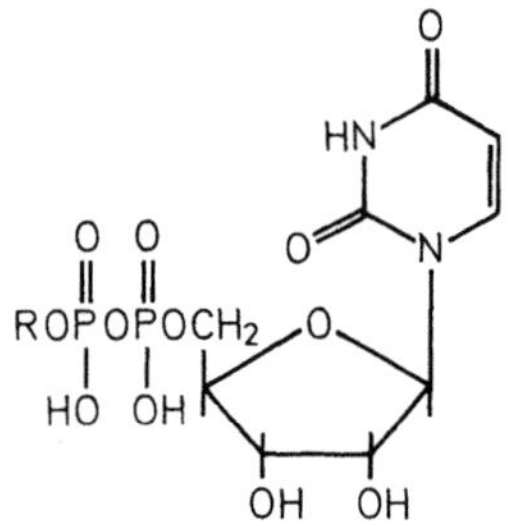

12 R = glucosyl
13 R = galactosyl
14 R = N-acetylo-glucosaminyl

The chains of atypical TAs are built by alternative addition of glycerol phosphate from CDP-glycerol and of glucose or galactose from UDP-glucose (*12*) or UDP-galactose (*13*), respectively [37]. The subsequent addition of these has been observed in the biosynthesis of the aTA from *Staphylococcus lactis* I3 (*4*). CDP-glycerol serves as a source of glycerol phosphate and D-N-acetylglucosamine-1-phosphate fragments come from UDP-GlcNAc (*14*) [38]. Both these phosphates are alternatively joined to the growing chain.

The *in vitro* [39] and *in vivo* [40] studies of *Streptococcus sanguis* allowed to establish that phosphatidylglycerol (*15*) serves as a precursor of the hydrophobic end of the LTA molecule as well as of the polyphosphate chain.

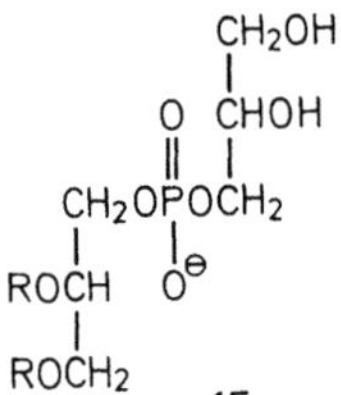

15
R = fatty acid residue

Recently, Childs and Neuhaus [41] described the *in vitro* acylation of polyglycerol phosphate chains with D-alanyl ester residues in a two-step reaction sequence. In the first step D-alanine is enzymatically transformed into an activated form, and in the second reaction the D-alanine molecule is incorporated into the LTA chain by a specific ligase.

5 Chemical Syntheses

There have only been a few papers published on the synthesis of simple TAs within the 25 years since their discovery.

5.1 Syntheses of TAs Fragments

Baddiley and coworkers [42] obtained 4-O-α-D-glucopyranosyl-D-ribitol (*18*), as the chain fragment of a TA extracted from *Lactobacillus arabinosus*. The synthesis involved several steps, as follows:

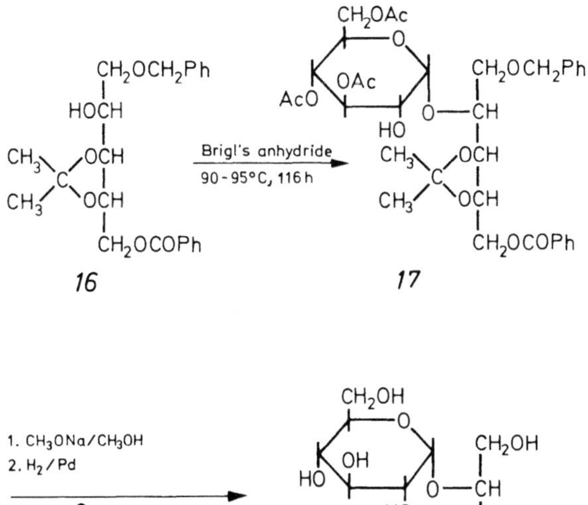

16 *17*

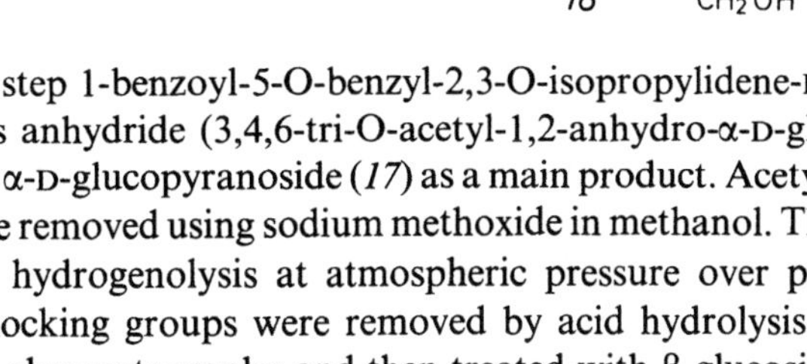

18

In the first step 1-benzoyl-5-O-benzyl-2,3-O-isopropylidene-D-ribitol (*16*) was heated with Brigl's anhydride (3,4,6-tri-O-acetyl-1,2-anhydro-α-D-glucose) yielding the corresponding α-D-glucopyranoside (*17*) as a main product. Acetyl and benzoyl protecting groups were removed using sodium methoxide in methanol. The benzyl ether bond was cleaved by hydrogenolysis at atmospheric pressure over palladium black. Isopropylidene blocking groups were removed by acid hydrolysis. Crude product *18* was purified by chromatography and then treated with β-glucosidase to cleave undesired β-glucoside. Final purification involved crystallization from ethanol. The pure product *18* showed the same properties (m.p., IR spectrum) as a glucoside from the *L. arabinosus* TA.

A similar approach was applied in the synthesis of α- and β-anomers of 4-O-D-glucosaminyl-D-ribitol *21*, the fragments of the TA extracted from *Staphylococcus aureus* H [43]. The sugar moieties were linked by using the free hydroxyl group of *16* and 3,4,6-tri-O-acetyl-N-4′-methoxybenzylidene-α-D-glucosamine bromide (*19*).

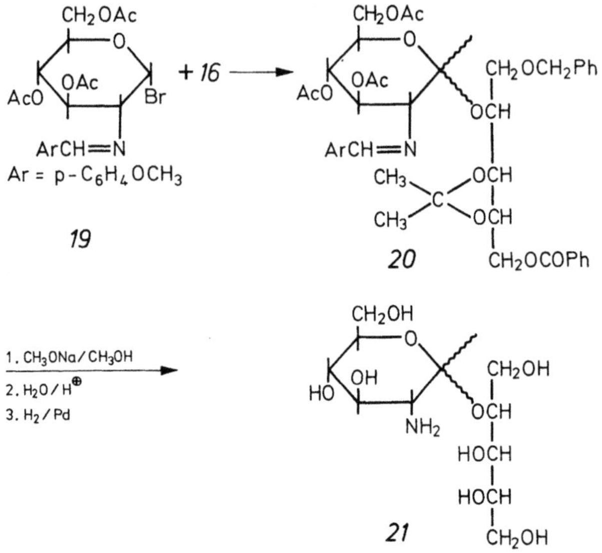

The first reaction yielded a mixture consisting of 88% of the α- and 12% of the β-anomer of *20*. The blocking groups were removed from *20* in three steps. Treatment of *20* with sodium methanolate in methanol allowed the acetyl and benzoyl groups to be removed. Benzylidene and isopropylidene protecting groups underwent acidic hydrolysis. The benzyl ether bond was cleaved by hydrogenation over palladium under atmospheric pressure. The resulting mixture of anomers of *21* was separated on Dowex 50 (H^+) using a 0.26 *N* HCl solution as eluent.

Both above preparation were described in the early sixties when instrumental methods of identification of chemical compounds were seldom used. The synthesis of the corresponding compounds was a confirmation of the postulated molecular structure of the natural products, and both described above syntheses were elaborated for the same reason.

Recently, the synthesis of 1-deoxy-3-O-phosphono-D-glycerol-1-yl-β-D-glucopyranoside disodium salt *27* and its L-glycerol-1-yl isomer *31* was described [45]. Compound *27* is a repeating unit of TA chains from *Bacillus subtilis* var. *niger* WM (formula *3*). Both *27* and *31* were obtained mainly to record their ^{13}C-NMR spectra to be further compared with the spectrum of the naturally occurring teichoic acid. Thus, *27* was prepared in a few-step synthesis as shown below:

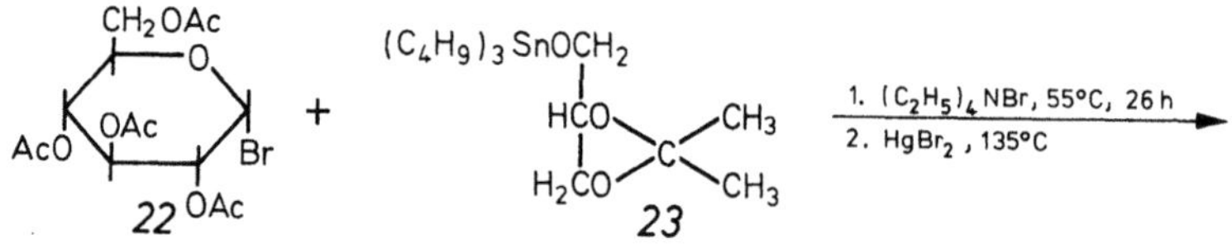

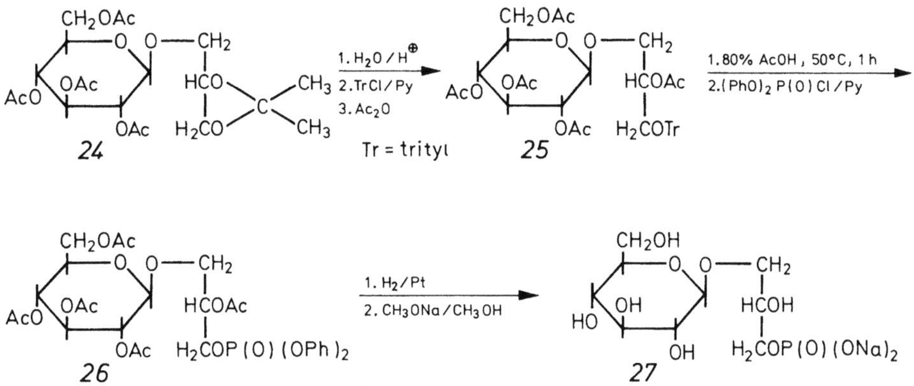

The β-glucoside *24* was prepared using the orthoester method in the condensation reaction of *22* with 2,3-O-isopropylidene-1-O-(tributyltin)-D-glycerol *23* followed by treatment of the resulting orthoester with mercuric bromide. The isopropylidene group of *24* was removed by acidic hydrolysis. Tritylation of the resulting β-glucoside followed by esterification with acetic anhydride gave trityl acetate derivative *25*. The latter, after detritylation in 80% acetic acid, gave a derivative with a free hydroxyl group at the 3 position, suitable for the phosphorylation reaction with diphenylphosphorochloridate in the presence of pyridine. Hydrogenolysis of the phenyl groups of *25* over platinum oxide and then deacylation with sodium methanolate in methanol gave *27*.

31 was prepared by inversion of the 2 carbon atom configuration in the glycerol moiety:

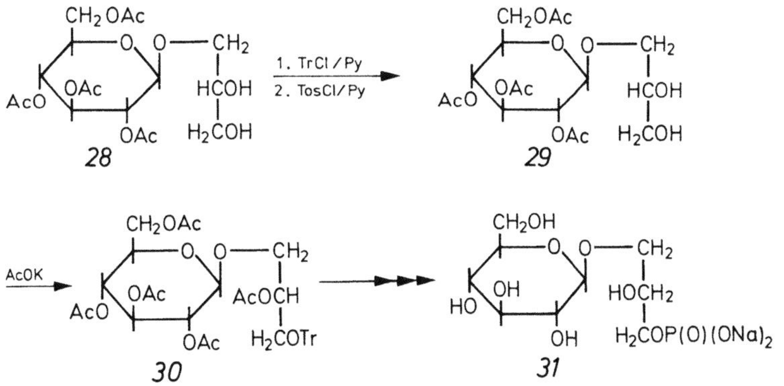

Tritylation and then tosylation of *28* afforded trityl tosylate derivative *29* suitable for epimerization at C-2. Epimerization was achieved using potassium acetate with dibenzo-18-crown-6. The trityl acetate *30* was transformed into *31* using the same route as described above for *25*. ^{13}C-NMR spectra of *27* and *31* are slightly different in chemical shift for C-2 signals due to the different configuration at C-2. The comparison of *27* and *31* with the natural sample confirmed the D-configuration of the glycerol moiety (*27*) in the naturally occurring TA. The bisphosphate *32* was obtained for ^{13}C-NMR studies starting with *28*:

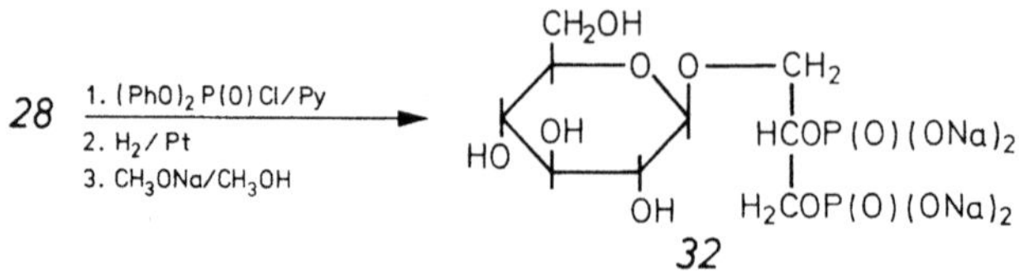

5.2 Chemical Syntheses of TAs Chains and their Models

5.2.1 Polycondensation Methods

Michelson first applied polycondensation (polyaddition) to prepare the TA-like backbones [1]. This method introduces the reactive pyrophosphate intermediate:

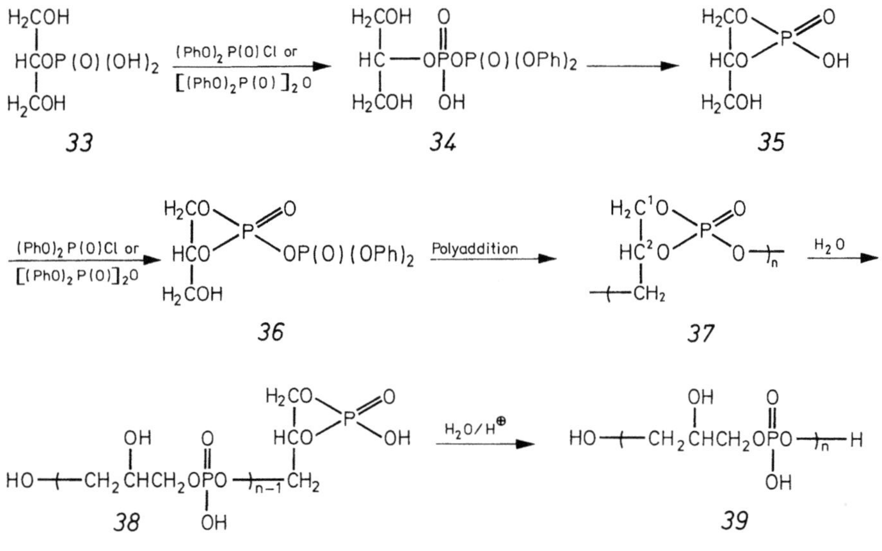

The cyclic glycerol phosphate *35* was obtained in the reaction of the mono-tri-*n*-octylammonium salt of glycerol 2-phosphate *33* with the phosphorylating agent in the presence of tri-*n*-butylamine. Phosphorylation of the free hydroxyl groups of glycerol moieties was not observed under the applied conditions. The phosphorylating agent first formed the pyrophosphate intermediate *36* (not isolated), which in turn gave polymer *37* in the polyaddition process conducted in dioxan solvent at 25 °C. The water hydrolysis opened one of the C—O bonds in the cyclic phosphate yielding a linear polymer *38* terminated with a cyclic end group. Treatment of *38* with diluted HCl opened the ring of the terminal cyclic phosphate. Finally, polymer *39* was isolated as a calcium salt and dialized several times against water and 2*N* NaCl. The average chain length of the purified *39* was equal to 12 structural units. This was determined by measuring the ratio of terminal phosphates to the total phosphate units. The terminal phosphate group was liberated from *39* using prostatic monoesterase and then determined as inorganic phosphate. *39* possessed some amount of incorrect 2,3-diester bonds in the glycerol moieties as a result of a cleavage of P—O—C^1 bonds during hydrolysis. In spite of this structural imperfection the Michelson's product had properties similar to the naturally occurring glycerol polyphosphate and reacted specifically with anti-polyglycerol phosphate antibodies [45].

In 1965, Baddiley and coworkers[2] described a preparation of the ribitol polyphosphate. This approach was based on the condensation of pyrophosphates:

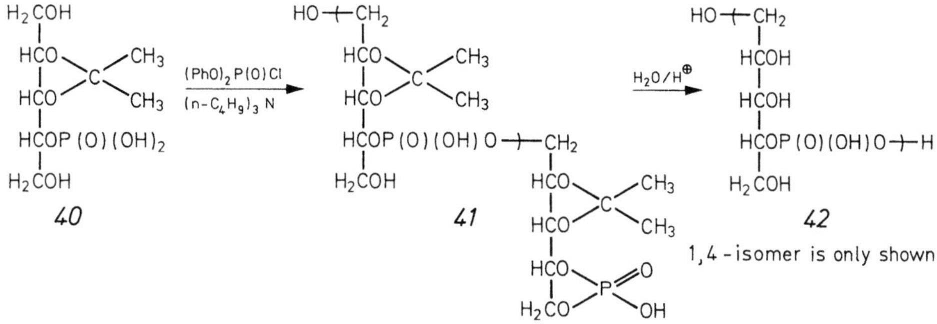

The mixture of monoesters (isomeric 5-phosphate is not shown) *40* was obtained by the phosphorylation of the 1,5-di-O-benzoyl-2,3-O-isopropylidene-D-ribitol with diphenyl phosphorochloridate followed by treatment with methanolic ammonia, ammonia, and finally aqueous barium hydroxide in order to remove benzoyl and phenyl blocking groups. The tri-*n*-octylammonium salts of *40*, treated with diphenyl phosphorochloridate, gave first the cyclic 4,5-phosphate (similar structure to *35*), which then gave the polymer *41*. The isopropylidene blocking groups were removed and the terminal cyclic phosphates were opened by diluted HCl. *42* was isolated in the form of the lithium salt and was purified by dialysis. The average $\overline{DP}_n = 5$ to 6. The final product contained mixed 1,4- and 1,5-phosphodiester linkages.

More recently, Van Boeckel et al. [46] described the synthesis of the chain fragment of a TA from *Bacillus subtilis* var. *niger* WM (formula *3*). This fragment consists of three units and was prepared by the step-by-step phosphotriester method. First the properly protected terminal unit *43* and middle unit *44* were condensed using 2,4,6-triisopropylbenzenesulfenyl-3-nitro-1,2,4-triazolide (TPSNT) as the condensing agent. The resulting dimer *45*, after purification by short column chromatography, was reacted with isopentyl nitrite in a pyridine/acetic acid mixture in order to obtain *46* with a free P—OH function. In the second step, dimer *46* was condensed with terminal unit *47* using TPSNT. Deblocking of trimer *48* required four steps: 1) the 2,2,2-tribromoethyl group was removed using zinc dust in pyridine with *p*-toluenesulfonic acid; 2) the 2-chlorophenyl group was removed with N^1,N^1,N^2,N^2-tetramethylguanidium-syn-4-nitro-benzaldoximate in dry THF; 3) the acetic ester bonds were hydrolyzed with aqueous ammonia. The hydrolysis product was purified by anion exchange chromatography and 4) finally, the removal of

MMTR and THP groups was performed by aqueous HCl. *49* was isolated as the ammonium salt. The structure of *49* was confirmed by ^1H, ^{13}C, and ^{31}P NMR.

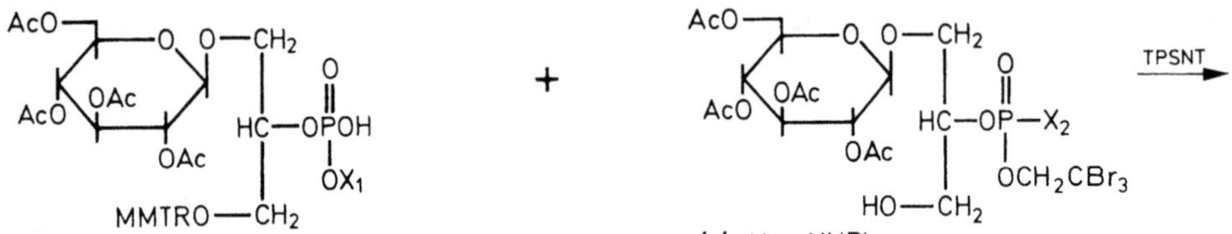

43 X_1 = o-ClPh
MMTR = monomethoxytrityl

44 X_2 = NHPh

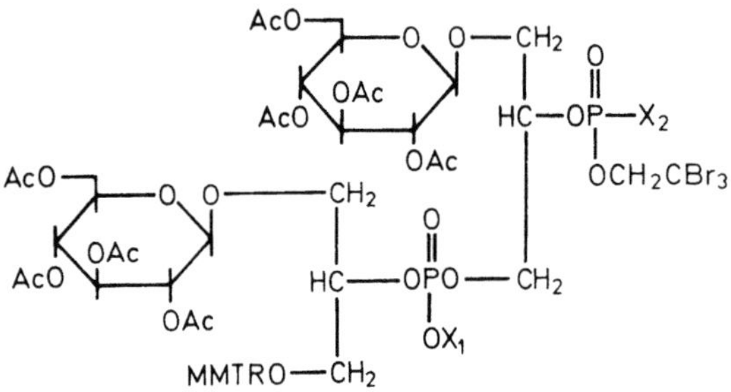

45 X_1 = o-ClPh, X_2 = NHPh
46 X_1 = o-ClPh, X_2 = OH

46 +

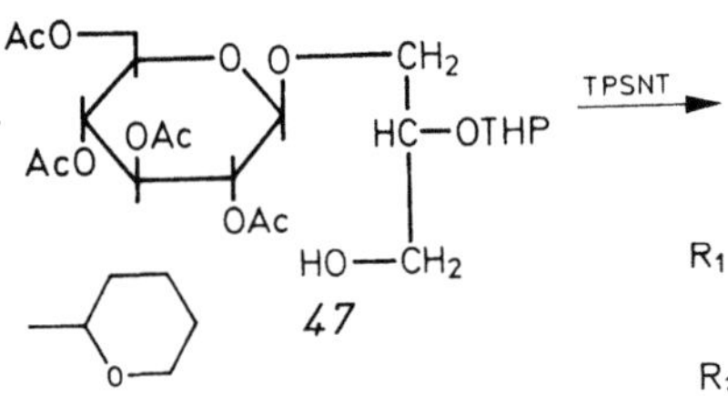

THP =

47

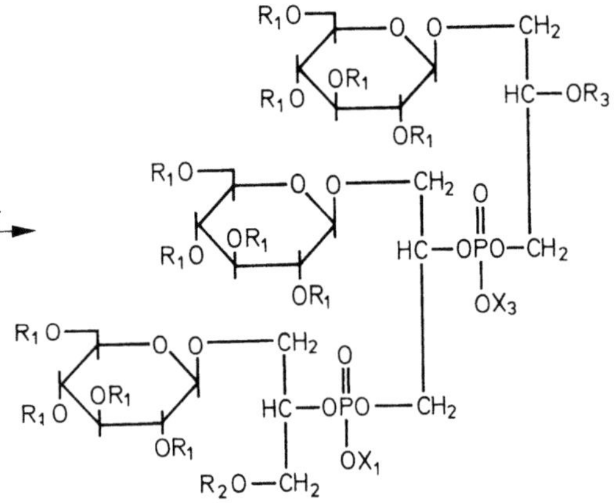

48 X_1 = o-ClPh, X_3 = CH_2CBr_3, R_1 = Ac, R_2 = MMTR, R_3 = THP
49 X_1 = X_3 = R_1 = R_2 = R_3 = H

5.2.2 Ring-Opening Polymerization Method

This method was already used in chemical syntheses of models of polypeptides [47] and polysaccharides [48].

To prepare models of TA, the corresponding cyclic alkylene phosphoric acids cannot be directly polymerized because the presence of acidic protons of the phosphate group makes the ionic polymerization impossible. Thus, cyclic phosphorus compounds with blocked third functions have to be used. These are phosphates, phosphoramidates and phosphites. After polymerization the obtained polymer is converted by deblocking into the polyalkylene phosphate, e.g. polysalt or polyacid form.

First, the simple high-molecular-weight polymer with the same sequence of atoms in the chain as in the glycerol TA was prepared in the following way [49]:

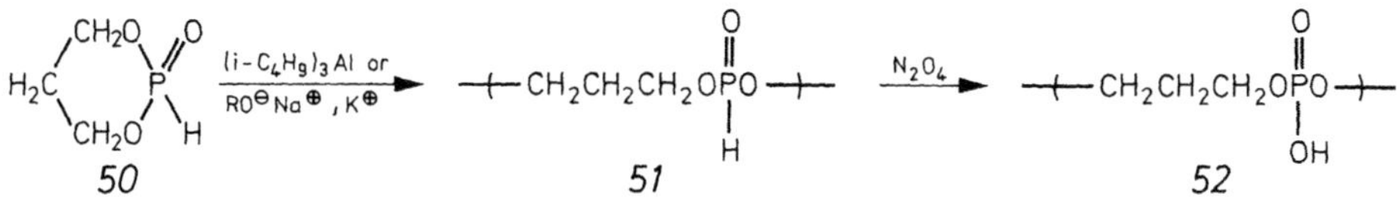

50 51 52

Cyclic phosphite *50* (prepared in high yield from diethylphosphite and 1,3-propyleneglycol) was polymerized at low temperature in the presence of $(i\text{-}C_4H_9)_3Al$ or metal alkoxides. The yield of *51* ranged, depending on temperature, from 50 to 75% (because of reaching the polymer-monomer equilibrium). The $\overline{M}_n$ (osmometry) was from 1.1×10^4 to 1.02×10^5 daltons [50]; polymerization can be performed as a living process. *51* had a regular linear structure confirmed by ^1H-, ^{31}P-, and ^{13}C-NMR spectra [50]. *51* could easily be oxidized with N_2O_4 in CH_2Cl_2 to the corresponding polyacid *52*. The oxidation proceeded quantitatively, and the final polyalkylenephosphate precipitated from the reaction mixture. The molecular weight measurements of *52* by light scattering indicated that during oxidation the disrupture of the chain was not important: about two broken bonds for over 10^3oxidized polymer units [51]. *52* is a colourless rubbery solid soluble in water. Polyacid *52* is stable in water for months, as it was indicated by the viscosity measurements.

Linear polyphosphite *51* was also obtained from a cyclic phosphorus-N,N-diethylamide *53* [52].

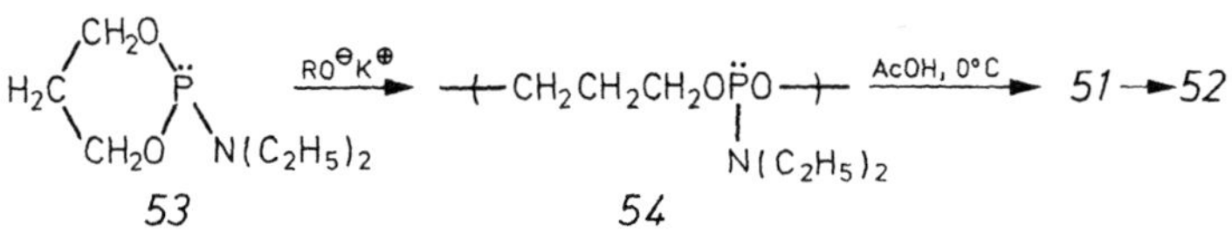

53 54

53 was polymerized in THF using potassium or cesium alcoholates as initiators. The sodium and lithium alcoholates were inactive, apparently because of their association, or association of the formed alcoholate active species. *54* was isolated and purified by several precipitations from THF into acetonitrile, in which *53* is soluble. $\overline{M}_n$ of purified *54*, measured by high speed osmometry, ranged from 5.0×10^3 to 2.5×10^4. *54* is a colourless, rubbery material soluble in THF, CH_2Cl_2, benzene and acetone. *54* could be easily converted into polyphosphite *51* using dry acetic acid. The progress of reaction can be monitored by $^{31}P\{^1H\}$NMR.

Early observations suggested [53] that cyclic six-membered phosphates only oligomerize because the polymerization process is accompanied by effective chain transfer to the monomer. This phenomenon was of little importance in the polymerization of much more strained five-membered cyclic phosphates *55*. Thus, polymerization of *55* yielded corresponding high-molecular-weight polymers *56* with a very good yield [54]:

R^1	R
H	CH_3
H	C_2H_5
H	n-C_3H_7
H	i-C_3H_7
H	CCl_3CH_2
H	n-C_4H_9
H	C_6H_5
CH_2Cl	C_2H_5

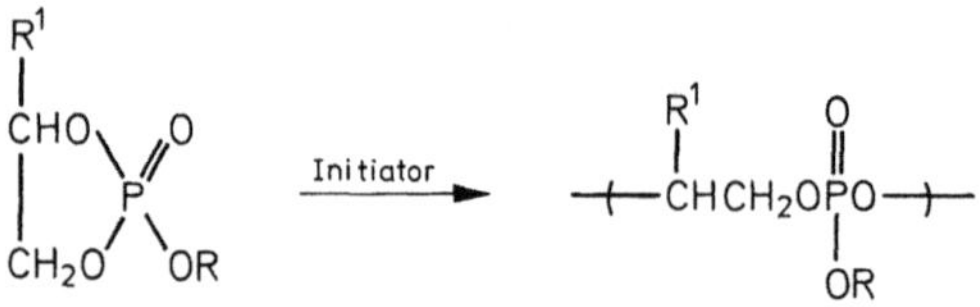

55 56

Polymerization proceeded at room temperature and was initiated by metal alkoxides or organometallic compounds (($(i\text{-}C_4H_9)_3Al$, $(C_2H_5)_2Zn$). $\overline{M}_n$ of these polymers was

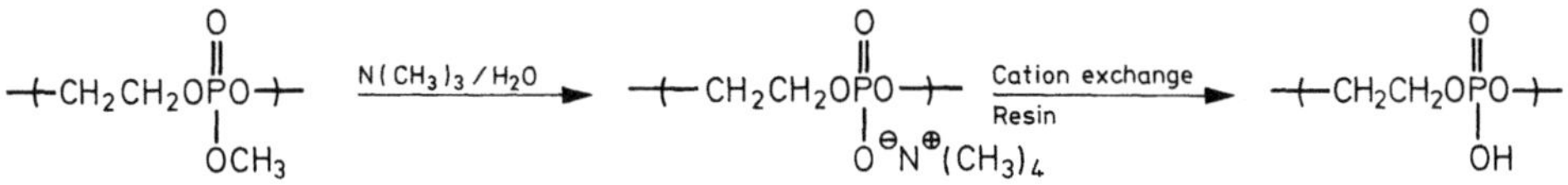

57 58 59

well above 10^4, sometimes 10^5. *57* was converted into polysalt form *58* by the dealkylation reaction with trimethylamine in water [49]. Polysalt *58* was passed through the cation exchange resin (Dowex 50, H^+ form) in order to obtain the polyphosphate (polyacid) *59* with backbones similar to those in glycerol TA with 2,3 linkages. The ^{1}H-NMR spectrum of *59* showed that the dealkylation process was not quantitative and about 10% of the methoxy groups remained unchanged on the polymer.

Similar results were obtained more recently by Yasuda et al. [55]. Monomers *60* and *61* were polymerized in benzene using $(C_2H_5)_2Mg$ as an initiator. The resulting *62* and *63* were insoluble in benzene. Both polymers heated in an Ar atmosphere at above 80 °C easily lost isobutylene and produced the required polyalkylenephosphates. The deblocking was almost quantitative (95% removal). $\overline{M}_n$ of polyacids *59* and *64*, measured by GPC, were 2.8×10^4 and 1.8×10^4, respectively.

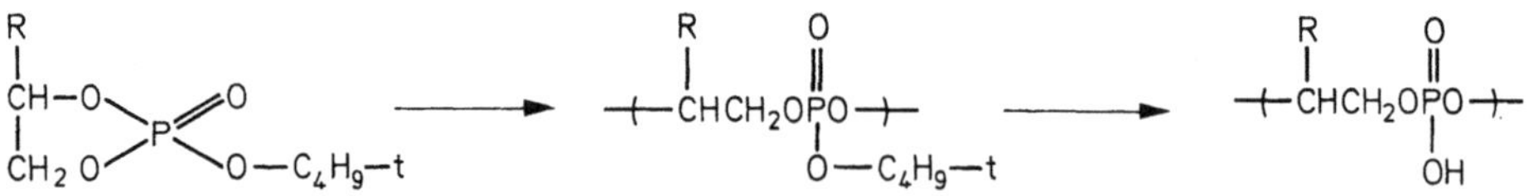

60 R = H
61 R = CH_3

62 R = H
63 R = CH_3

59 R = H
64 R = CH_3

Polymerization of 2-*t*-butoxy-4-methyl-2-oxo-1,3,2-dioxaphosphorinane *65* (a mixture of geometrical isomers) gave the high-molecular-weight polymer *66*. Polymerization was carried out in benzene at 40 °C using 1:0.9 $(C_2H_5)_3Al:H_2O$ as a catalyst. The resulting polymer *66* was heated at 130 °C for 20 min to obtain the corresponding polyacid *67* with a molecular weight of 2.5×10^4 (GPC).

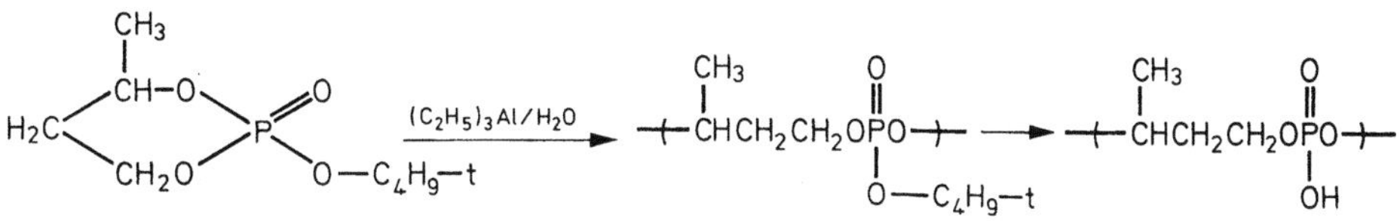

65 66 67

Polymerization of the properly substituted, cyclic five-membered monomers allowed us to prepare models of glycerol TA with 2,3-phosphodiester linkages [56]. Both approaches described previously were examined. The mixture of diastereoisomers of the cyclic phosphate (*55*, R = CH_3, R^1 = $AcOCH_2$) was polymerized in CH_2Cl_2 using $(i\text{-}C_4H_9)_3Al$ as an initiator to yield a polymer with a molecular weight of up to 2.5×10^4 (high speed osmometry). The quantitative conversion of the corresponding polyester (*56*, R = CH_3, R^1 = $AcOCH_2$) into the polysalt form could not be achieved. The best result was obtained using NaJ in acetone as a dealkylating agent: more than 75% dealkylation was observed. The second method was based on the phosphite route, similar to preparation of *52*. The monomer, the corresponding cyclic alkylene phosphite *70*, was obtained starting from racemic glycerol α-monoacetate *68*.

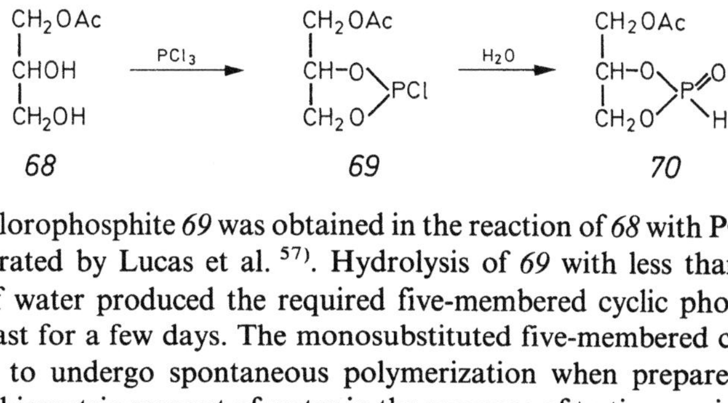

68 69 70

Cyclic chlorophosphite *69* was obtained in the reaction of *68* with PCl_3 under conditions elaborated by Lucas et al. [57]. Hydrolysis of *69* with less than stoichiometric amounts of water produced the required five-membered cyclic phosphite *70*, being stable at least for a few days. The monosubstituted five-membered cyclic phosphites are known to undergo spontaneous polymerization when prepared by hydrolysis with a stoichiometric amount of water in the presence of tertiary amines used for HCl traping [58,59].

Polymerization of a mixture of diastereoisomers of phosphite *70* with $(i\text{-}C_4H_9)_3Al$ in CH_2Cl_2 solution at room temperature yielded the corresponding polyphosphite *71*, which was subsequently oxidized with N_2O_4 without preliminary isolation and purification:

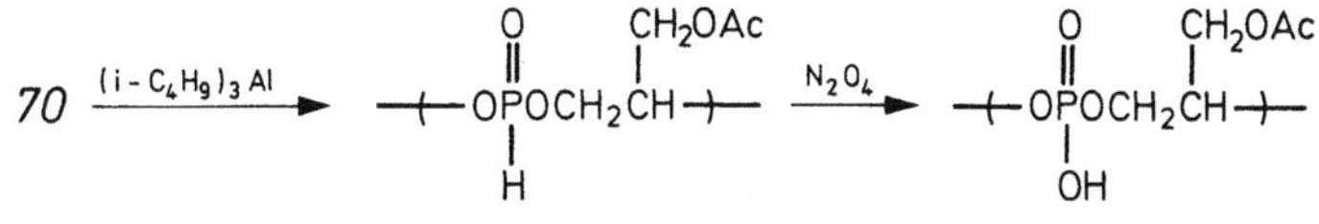

71 72

The polyacid *72* was isolated and purified by precipitation with an overall yield of about 50%. *72* is a colourless, brittle material soluble in DMF, DMSO and water and unstable in aqueous solutions. In water, *72* lost acetic acid ester groups. Similar lability of the ester bonds at pH ≠ 7 was observed for natural D-alanyl-substituted LTA [60]. However, *72* neutralized with Na_2CO_3 yielded a polymeric sodium salt, which was stable in water [61]. The $\overline{M}_n$ of *72*, measured by vpo in DMF, reached 1.0 × 10⁴ (n = 50). Light scattering measurements in H_2O for the sodium salt of *72* gave a $\overline{M}_w$ of 6.7×10^4. ^{1}H- and ^{13}C-NMR spectra of *72* are given in Fig. 1 and confirmed the structure of the monomeric unit of *72*. $^{31}P\{^1H\}$NMR spectrum of *72* in DMSO consisted of one signal at δ = −2.3 ppm, suggesting a regular head-to-tail structure. However, the $^{31}P\{^1H\}$NMR spectrum of the sodium salt of *72* in D_2O revealed three signals at δ = 0.2 ppm, δ = −0.56 ppm, and δ = −1.09 ppm with intensities of 0.5:1:0.5, respectively [61]. This fact may suggest that the ring opening of the cyclic phosphite *70* under the studied conditions involves two pathways with similar probability, leading to a linear chain possessing diads with head-to-head, head-to-tail, and tail-to-tail units. A similar phenomenon was observed in the polymerization of 4-methyl-2-hydro-2-oxo-1,3,2-dioxaphospholane [62].

Treatment of *72* with methanolic ammonia yielded ammonium polysalt *73* with free hydroxyl groups.

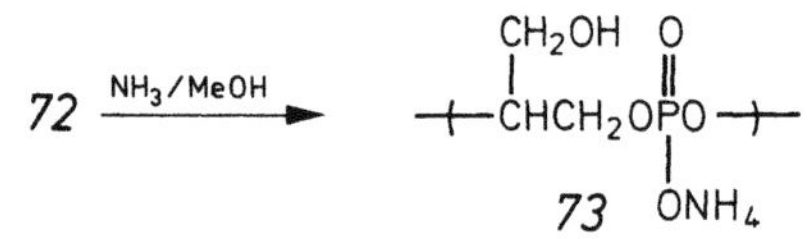

The removal of acetic acid ester groups was quantitative as it was shown by the ^{1}H- and ^{13}C-NMR spectra [61], given in Fig. 2. The $^{31}P\{^1H\}$NMR spectrum of *73* consists of three peaks at δ = 0.85, 0.18 and −0.37 ppm.

Gehrmann and Vogt [63] described the polymerization of a bicyclic phosphate *74*

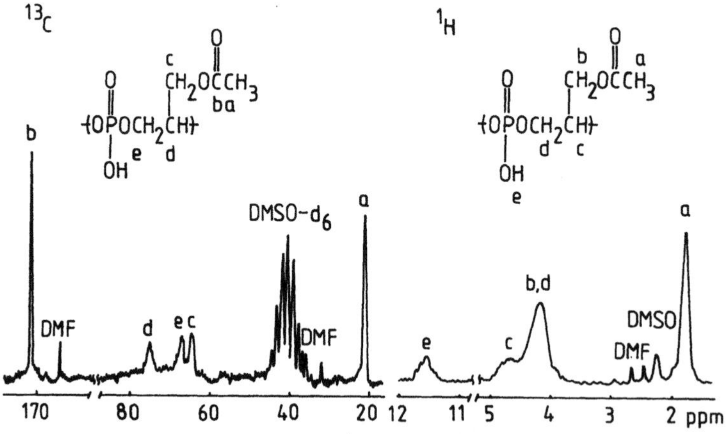

Fig. 1. $^{13}C\{^1H\}$ and ^{1}H NMR spectra of *72* in CD_3SOCD_3 (30% solution, 25 °C)

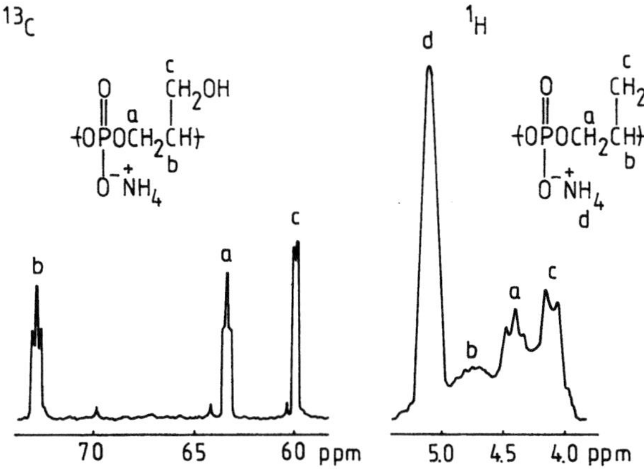

Fig. 2. $^{13}C\{^1H\}$ and ^1H NMR spectra of *73* in D_2O (30% solution, 25 °C)

to the corresponding crosslinked polymer *75*. *74* is a very reactive monomer and could be cationically polymerized at −78 °C with traces of water as an initiator:

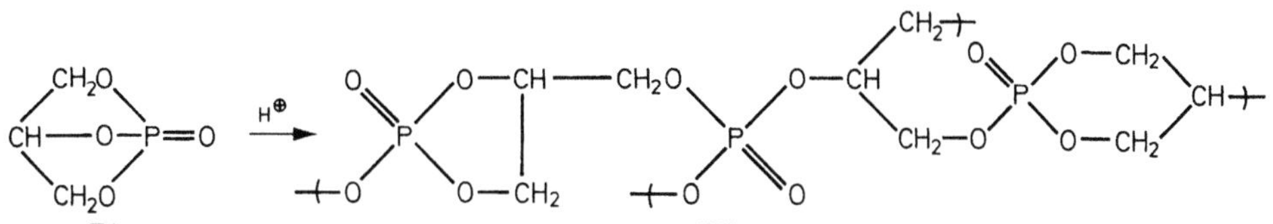

74 *75*

75 was hydrolyzed under controlled conditions to give linear polymer *76* with a $\overline{M}_n$ of up to 9.0×10^3.

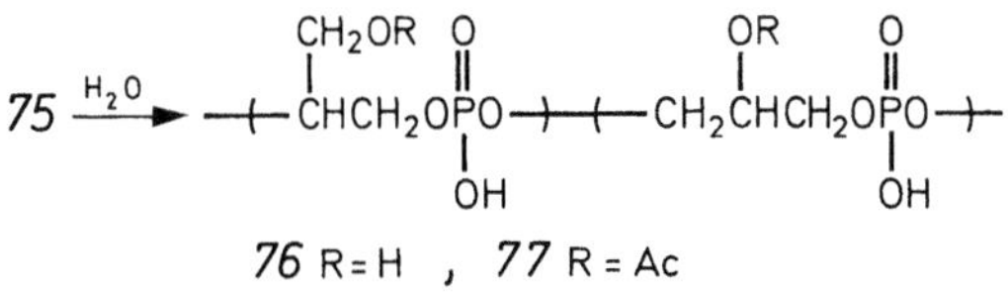

76 R = H , *77* R = Ac

The linear polymer *76* consisted of both kinds of repeating units, namely with 1,3- and 2,3-linkages in the glycerol moieties. Acylation of *76* with acetic anhydride in pyridine yielded *77*.

6 References

1. Michelson, A. M.: J. Chem. Soc. *1959*, 1371
2. Applegarth, D. A., Buchanan, J. G., Baddiley, J.: J. Chem. Soc. *1955*, 1213
3. Penczek, S., in: "Phosphorus Chemistry Directed Towards Biology"; Stec, W. J. Ed., Pergamon Press: Oxford and New York, *1980*, 133
4. Baddiley, J.: Eassay Biochem. *8*, 35 (1972)
5. Lambert, P. A., Hancock, J. C., Baddiley, J.: Biochem. Biophys. Acta *472*, 1 (1977)
6. Wicken, A. J., Knox, K. W.: Science *187*, 1161 (1975)

7. Dziarski, R.: Curr. Top. Microbiol. Immunol. *76*, 113 (1976)
8. Baddiley, J., Buchanan, J. G., Martin, R. O., RajBhandary, H. L.: Biochem. J. *85*, 49 (1962)
9. a) Davison, A. L., Baddiley, J.: Gen. Microbiol. *32*, 271 (1963);
   b) Sharpe, M. E., Davison, A. L., Baddiley, J.: J. Gen. Microbiol. *34*, 333 (1964)
10. Burger, M. M.: Proc. Nat. Acad. Sci. N. S. *56*, 910 (1966)
11. Torii, M., Kabat, E. A.. Bezer, A. E.: J. Expt. Med. *120*, 13 (1964)
12. De Boer, W. R., Kruyssen, F. J., Wouters, J. T. M., Kruk, C.: Eur. J. Biochem. *62*, 1 (1976)
13. a) Dmitrieva, N. F., Naumova, J. B., Panina, L. J.: Biokhimiya *40*, 211 (1975);
    b) Streshinskaya, G. M., Naumova, J. B., Panina, L. J.: Mikrobiologiya *48*, 814 (1979)
14. Naumova, J. B., Shashkov, A. S., Scoblilova, N. K., Agre, N. S., Romanov, V. V.: Bioorg. Khim. *8*, 848 (1982)
15. a) Fiedler, F., Schäffler, M. J., Stackenbrandt, E.: Arch. Microbiol. *129*, 85 (1981);
    b) Voerkamp, J. H., Hoelen, G. E. J. M., Op Den Camp, H. J. M.: Biochim. Biophys. Acta *755*, 439 (1983)
16. Archibald, A. R., Baddiley, J., Heckels, J. E., Heptinstall, S.: Biochem. J. *125*, 353 (1971)
17. De Boer, W. R., Wouters, J. T. M., Anderson, A. J., Archibald, A. R.: Eur. J. Biochem. *85*, 433 (1978)
18. Branefors-Helander, P., Erbing, Ch., Kenne, L., Lindberg, B.: Carbohydr. Res. *56*, 117 (1977)
19. Watson, M. J., Baddiley, J.: Biochem. J. *137*, 399 (1974)
20. Watson, M. J., Tyler, J. M., Buchanan, J. G., Baddiley, J.: Biochem. J. *130*, 45 (1972)
21. Egan, W., Schneerson, R., Werner, E. K., Zon, G.: J. Am. Chem. Soc. *104*, 2898 (1982)
22. Archibald, A. R., Stafford, G. H.: Biochem. J. *130*, 681 (1972)
23. Arakawa, H., Shimada, A., Ishimoto, N., Ito, E.: J. Biochem. (Tokyo) *89*, 1555 (1981)
24. Toon, P., Brown, P. E., Baddiley, J.: Biochem. J. *127*, 399 (1972)
25. Button, D., Hennings, N. L.: J. Bacteriol. *128*, 149 (1976)
26. Nakano, M., Fischer, W.: Hoppe-Seyler's Z. Physiol. Chem. *359*, 1 (1978)
27. Fischer, W., Lame, R. A., Nakano, M.: Biochim. Biophys. Acta *528*, 298 (1978)
28. Baddiley, J., Hancock, J. C., Sherwood, P. M. A.: Nature *243*, 43 (1973)
29. Lambert, P. A., Hancock, I. C., Baddiley, J.: Biochem. J. *149*, 519 (1975)
30. Heckels, J. E., Lambert, P. A., Baddiley, J.: Biochem. J. *162*, 359 (1977)
31. Hughes, A. H., Hancock, J. C., Baddiley, J.: Biochem. J. *132*, 83 (1973)
32. Herbodl, D. R., Glaser, L.: J. Biol. Chem. *250*, 7231 (1975)
33. Burger, M. M., Glaser, L.: J. Biol. Chem. *239*, 3178 (1964)
34. Glaser, L.: J. Biol. Chem. *239*, 3178 (1964)
35. Glaser, L., Burger, M. M.: J. Biol. Chem. *239*, 3187 (1964)
36. Ishimoto, N., Strominger, J. L.: J. Biol. Chem. *241*, 639 (1966)
37. Burger, M. M., Glaser, L.: J. Biol. Chem. *241*, 494 (1966)
38. Baddiley, J., Blumson, N. L., Douglas, L. J.: Biochem. J. *110*, 565 (1968)
39. Emdur, L. J., Chiu, T. H.: FEBS Lett. *55*, 216 (1975)
40. Emdur, L. J., Chiu, T. H.: Biochem. Biophys. Res. Commun. *59*, 1137 (1974)
41. Childs, W. C., Neuhaus, F. C.: J. Bacteriol. *143*, 293 (1980)
42. Sargent, L. J., Buchanan, J. G., Baddiley, J.: J. Chem. Soc. *1962*, 2184-7
43. Hardy, F. E., Buchanan, J. G., Baddiley, J.: J. Chem. Soc. *1963*, 3360-6
44. Ogana, T., Katano, K., Matsui, M.: Carbohydr. Res. *70*, 37 (1979)
45. McCarty, M.: J. Exp. Med. *109*, 361 (1959)
46. Van Boeckel, C. A. A., Visser, G. M., Hermans, J. P. G., Van Boom, J. H.: Recl. Trav. Chim. Pays. Bas *102*, 526 (1983)
47. Imanishi, Y., in: "Ring-Opening Polymerization"; Ivin, K. I. and Saegusa, T. Ed., Elsevier: London and New York, *1984*, 523
48. Schuerch, C.: Acc. Chem. Res. *6*, 184 (1973)
49. Kałużyński, K., Libiszowski, J., Penczek, S.: Macromolecules *9*, 365 (1976)
50. Kałużyński, K., Libiszowski, J., Penczek, S.: Makromol. Chem. *178*, 2943 (1977)
51. Kałużyński, K.: unpublished results
52. Pretula, J., Kałużyński, K., Penczek, S.: J. Polymer Sci., Polymer Chem. Ed. *22*, 1251 (1984)
53. Łapienis, G., Penczek, S.: Macromolecules *10*, 1301 (1977)
54. Libiszowski, J., Kałużyński, K., Penczek, S.: J. Polymer Sci., Polymer Chem. Ed. *16*, 1275 (1978)
55. Yasuda, H., Sumitani, M., Nakamura, A.: Macromolecules *14*, 460 (1981)

56. Kłosiński, P., Penczek, S.: Macromolecules *16*, 316 (1983)
57. Lucas, H. J., Mitchell, F. W., Scully, C. N.: J. Am. Chem. Soc. *72*, 5491 (1950)
58. Zwierzak, A.: Can. J. Chem. *45*, 2501 (1967)
59. Nifant'ev, E. E., Nasonowskij, J. S., Borisenko, A. A.: J. Gen. Chem. (USSR) *41*, 1876 (1971)
60. Childs III, W. C., Neuhaus, F. C.: J. Bacteriol. *143*, 293 (1980)
61. Kłosiński, P., Penczek, S.: in preparation
62. Biela, T., Penczek, S.: in preparation
63. Gehrmann, T., Vogt, W.: Makromol. Chem. *182*, 3069 (1981)

Editor: C. G. Overberger
Received January 17, 1986

Analysis of Compositional and Structural Heterogeneities of Polymers by Non-Exclusion HPLC

Gottfried Glöckner
Technische Universität Dresden, Sektion Chemie
Mommsenstr. 13, DDR-8027 Dresden/GDR

High performance liquid chromatography (HPLC) of polymers is often thought to be synonymous with Size Exclusion Chromatography for separation by molar size. The present article deals with non-exclusion chromatography of polymers which enables separation by differences in the chemical structure and composition.

Abbreviations

AN	Acrylonitrile
AC	Adsorption chromatography
AcN	Acetonitrile
CF	Cross fractionation
CCF	Chromatographic cross-fractionation
CCD	Chemical-composition distribution
HIC	Hydrophobic interaction chromatography
IEC	Ion exchange chromatography
MEK	Methyl ethyl ketone
MMD	Molar-mass distribution
RP	Reversed phase
RPC	Reversed-phase chromatography
PS	Polystyrene
SAN	Poly(styrene-*co*-acrylonitrile)
SMA	Poly(styrene-*co*-methyl acrylate)
SMMA	Poly(styrene-*co*-methyl methacrylate)
SEC	Size exclusion chromatography
SBR	Styrene-butadiene rubber
THF	Tetrahydrofuran
TLC	Thin-layer chromatography

Symbols

d_O	Pore diameter
d_P	Particle size of packing materials
P	Degree of polymerization
M	Molar mass
$\overline{M}_n$	Number average of M

The dimensions of columns are given as length × inner diameter d_O; d_P).
Gradient indication: A/B (B is the stronger solvent whose concentration increases in the course of the program)
Eluent mixtures of constant concentration are indicated by A—B

1 Introduction

1.1 Special Features of Polymers in Liquid Chromatography

"The difference between macromolecular and small-molecule chromatography, from a fundamental point of view, is considerably greater than the difference between gas and liquid chromatography".

J. C. Giddings [1)]

1.1.1 Particle Size to Pore Diameter

Most of the common packing materials for liquid chromatography are porous, usually with 6 nm pore diameter. The particle size of normal low-molecular substances is in the order of 1 nm, i.e., a high ratio of pore diameter to solute dimensions is characteristic of small-molecule chromatography. Polymer size can be characterized by coil dimensions, e.g., by the radius of gyration, or in the case of polymers with a tertiary structure, by the length of three chracteristic axes. Above 10,000 g/mol, the threshold of polymer behavior, the dimensions of molecules are equal to or larger than the pores of common packings. Polymer chromatography can be subdivided into size exclusion chromatography and separation by non-exclusion mechanisms. The rate of solute to pore size and the energy of interaction between the solute and the surface of the column packing decide whether exclusion would predominate or the non-exclusion phenomena. For non-exclusion chromatography wide-pore packings are recommended. The pores should be spacious enough to give even the largest solute molecules access to the active inner surface. Interference of non-exclusion and exclusion mechanisms is in most cases detrimental to straightforward separation strategies. On the other hand, the pore diameter should not be too large because this would reduce the surface area and thus the chromatographic activity of the packing.

Large molecules that just fit into the pores can contact relatively more active sites around the wall than small molecules which only touch the wall with one side of their surface. Experimental evidence for this idea can be derived from the adsorption of poly(methyl methacrylate) on silica with or without pores: the fraction of CO groups involved in adsorption was 0.32–0.39 on nonporous but 0.46–0.49 on porous silica [2)].

1.1.2 Effect of Interlinked Groups

A characteristic of small-molecule liquid chromatography is the reversibility of their contacts with the stationary phase. The distribution equilibrium constant determines the duration of the stationary periods and, thus, the retention of the solute. With polymers, isocratic retention factors of normal degree (i.e., $1 \leqq k' \leqq 10$) generally do not occur. A fractional alteration of elution conditions may cause transition from zero retention to infinity. As a rule of thumb, polymers either pass without retention or remain in the column. This "off or on" behavior produces the impression of irreversible fixation under the conditions of retention.

The principles of chromatographic retention are identical for polymers and low-molecular solutes, of course. Every adsorbed repeat unit of a polymer is reversibly attached to the stationary phase — but adjacent units are linked to each other through covalencies. The probability is small that all units of an adsorbed train can desorb simultaneously. Ignoring cooperative effects, let us attribute the inverse of the chromatographic distribution constant K^+ to the probability of desorption for a repeat unit. By definition, K^+ is the ratio of the concentrations in the stationary and mobile phase:

$$K^+ = c_{stat}/c_{mobil} \tag{1}$$

With the help of the phase ratio q, K^+ can be related to the time that a solute will stay in either the stationary or the mobile phase, t″ or t′, respectively:

$$K^+q = t''/t' = m''/m' = k' \tag{2}$$

The quantity k′ is the *retention factor* of the solute, or its mass distribution constant. It allows the straightforward comparison of results obtained with different apparatus and is frequently used in chromatographic literature. The optimum value is $k' = 2$ in isocratic elution. Here, on an average, the solute remains in the stationary phase twice as long as in the mobile phase. The probability of finding the solute in the mobile phase is $1:(1 + 2) = 0.33$.

A retention factor $k' = 2$ for a repeat unit of a polymer chain implies also the probability of desorption to be 0.33. If two units are linked and no cooperative interaction occurs, the probability of *simultaneous* desorption reduces to $0.33 \times 0.33 = 0.11$. (The likelihood of simultaneous adsorption is $0.67 \times 0.67 = 0.44$ and that of adsorption of one or the other unit is $2 \times (0.67 \times 0.33) = 0.44$.)

A molecule with several independently adsorbed units in a train can migrate only if all of them are free. Even for the simplest case of a "two-unit polymer" the k′ value becomes $k' = 0.88:0.11 = 8$ under chromatographic conditions yielding $k' = 2$ for an independent unit. For the adsorption of n units with a constant k'_u for each unit the retention factor for the whole train becomes:

$$k'_{total} = (k'_u + 1)^n - 1 \tag{3}$$

With real polymers, n is rather large. Hence, k'_u values only slightly exceeding zero give rise to high k'_{total} data. For example, $k'_{total} = 130$ for $k'_u = 0.05$ and $n = 100$. The logarithmic form of Eq. (3):

$$\log (k'_{total} + 1) = n \log (k'_u + 1) \tag{3'}$$

gives a relation like the well-known Martin rule for the members of a homologous series with carbon number n and a group-specific contribution A:

$$\log k'_{Martin} = A + Bn \tag{4}$$

The linearity proposed by Eq. (4) has been proven for a great variety of homologs but is no longer fulfilled by the higher members of the series. Here, the increase in log k′ with n declines. Recent work with polystyrene oligomers showed adherence to the Martin equation for solutes with no more than 19 repeat units, but log k′ still increasing with molar mass around 50,000 g/mol [3].

In binary eluents, k′ depends on the composition of the mobile phase. This dependence is approximately given through:

$$\log k' = \log k'_w - S\varphi \tag{5}$$

for reversed-phase chromatography with aqueous eluents containing the volume fraction φ of organic modifier. In Eq. (5), k'_w is the retention factor in pure water. A slope factor S in the order 1–5 is normal for low-molecular solutes and enables gradient elution to be performed within a reasonable interval of φ.

Equation (5) can be rewritten as:

$$(k'_{w,u}/k'_u) = 10^{S_u \varphi} \tag{5'}$$

The index ($_u$) indicates reference to one unit. Since $(k'_u + 1)^n \gg 1$ the combination of Eqs. (5′) and (3) for polymers with n units gives:

$$S_{total} = nS_u \tag{6}$$

The linear increase of S with n was established experimentally with polystyrene oligomers [3,4]. Of course, Eq. (6) is bound to the same decline at high degree of polymerization as Martin's rule but, at any rate, high polymers have high S values. They are in general so large that elution is restricted to an extremely narrow interval of eluent composition, *vide supra.*

1.1.3 Conformational Changes

Only some types of macromolecules have rigid structure. Among them are the native proteins with their secondary and tertiary structure stabilized by disulfide bridges, hydrogen bondings, and hydrophobic interactions. In general, random coil conformations are typical of polymers. In dilute solutions, these coils contain much solvent (about 99%). The polymer segments are movable. Special conformations may be favored by external forces. Under the conditions of high-pressure liquid chromatography, random coils may be deformed in a way that extended trains of segments come in close contact with the surface of the stationary phase and only a few loops and free chain ends protract into the mobile phase. In pores, coiled macromolecules may adjust their conformation to the size of the cavity in order to gain the maximum of interaction enthalpy. The flexibility can cause dramatic differences in chromatographic behavior by comparatively small changes in eluting conditions. An example for the influence of conformation on retention was recently found in reversed-phase chromatography of proteins under conditions of coexistence of native and unfolded structures [5].

1.1.4 Solubility

The thermodynamic condition for solubility is the negative change in Gibbs' free energy. With low-molecular solutes this is easily fulfilled because of the large entropy contribution. Since this quantity refers to the number of particles, it is rather small for polymer solutes. Here, the enthalpy contribution ΔH gains importance. Expressed in terms of Hildebrand's solubility concept, it means that a polymer can dissolve only in fluids whose solubility parameters are very close to that of the polymer. Therefore, the solvents available for a given polymer are far outnumbered by nonsolvents.

Increasing temperature widens the solubility window (i.e., the largest distance between the solubility parameter values of two solvents for a given polymer). This can be perceived from the Gibbs free energy equation. The same holds true for the narrowing effect of increasing molar mass that diminishes the entropy term.

1.2 General Remarks on Liquid Chromatography

1.2.1 Effect of Flow Rate and Particle Size

The efficiency of chromatographic systems can be defined by the height equivalent to a theoretical plate (HETP or plate height, symbol H) or by the reduced plate height:

$$h = H/d_P \tag{7}$$

with the particle diameter d_P. The reduced plate height is a function of the reduced velocity ν and can be expressed by the Knox equation:

$$h = B/\nu + A\nu^{1/3} + C\nu \tag{8}$$

with

$$\nu = ud_P/D \tag{9}$$

in which u is the linear velocity of the eluent in the column, and D the diffusion constant of the solute in the mobile phase. For a given system, A, B, and C are constants depending on homogeneity and density of the packing, longitudinal diffusion, and mass transfer, respectively.

According to Eq. (8), h has a minimum at low ν values corresponding to the optimum efficiency of the system. In polymer LC, the minimum of the h vs ν curve has not been realized yet because polymer solutes have diffusion constants which are smaller than those of low-molecular solutes by a factor of about 1000. Reasonable flow rates cause working conditions at the rising right-hand branch of the h vs ν curve. Thus, three consequences are obvious:

- Small-particle packings (d_P 4 or 3 μm) are recommended.
- High temperatures promise advantages through faster diffusion.
- The rate of flow should be as small as possible.

In SEC, even the last statement is definitely an improvement. Here, the ideal column packing is inert and the separation is solely due to entropy-governed distribution of the macromolecules between interparticle channels and pores. In separation mechanisms including energetic interactions between the surface of the stationary phase and flexible macromolecules, a steep increase of the reduced plate height was observed at very low flow rate that cancelled the improvement by flow reduction [6].

At a superficial glance, it could have appeared like the behavior predicted by Eq. (8), but the minimum occurred at far too high values of reduced velocity which increased with column length and the molar mass of the samples. With the increase of plate height at very low velocities the peaks became tailed. Thus it was concluded that the surprising effect was caused by a slow unfolding process of polymer coils, possibly to be interpreted as partial creeping of the flexible macromolecules into pores which are too small for accessibility by free penetration [6].

1.2.2 Influence of solvent strength

If there is no energy gain through the forming of contacts between the solute and the stationary phase, the distribution of solute molecules between the solvent inside and outside the pores will be governed by the change in entropy. If the pores are large enough to accommodate the polymer coils, the distribution of the latter between the interparticle volume and the pores yields a growth in entropy. Smaller pores would require some ordering of the flexible chains and hence a decrease in entropy. Thus, size exclusion chromatography is an entropy-controlled separation where retention decreases with increasing size of the solute. The well-known representation of this behavior is the slanting curve of log M vs V_e (see Fig. 1, line "100").

If the chromatographic strength of the solvent diminishes beyond a critical value, the macromolecules will interact with the surface. The gain in enthalpy is proportional to the number of segments involved, hence, retention increases with molar mass (see Fig. 1, lines "5" or "0"). The enthalpic interaction is strongly dependent on the chemical structure of both the surface and the solute.

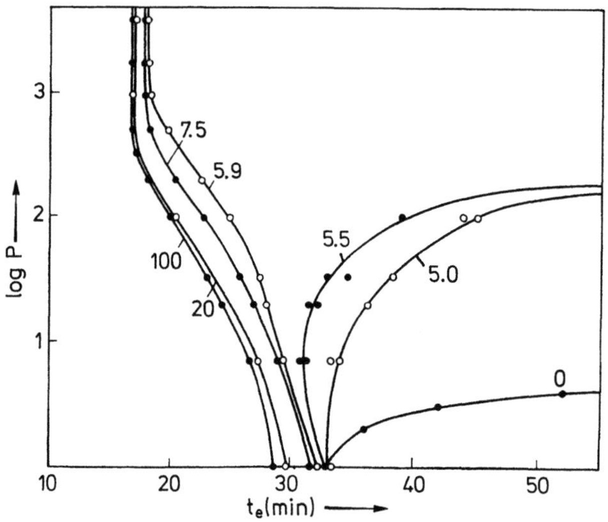

Fig. 1. Solvent-strength effect. Elution characteristics of polystyrene standards from a silica column (600 × 4 mm; d_o = 10 nm) at 30 °C. Eluent: tetrachloromethane — chloroform mixtures (the figures indicate the chloroform content, vol %). P — degree of polymerization. (From Ref. [7] with permission)

Obviously, the enthalpy gain can compensate for some unfavorable change in entropy. At a suitable value of adsorption energy, flexible macromolecules may reptate into narrow pores. Polystyrene of molar mass 173,000 g/mol and coil diameter 30 nm is reported to be capable of entering pores of silica gel with 10 nm average pore diameter [8]. This penetration will certainly be a slow process. It often has been observed that the amount of polymer adsorbed by porous adsorbents slowly increases over a protracted period of time. This may be due to similar effects.

Figure 1 was the first published information about the behavior of polymers in eluents graded in elution strength [7]. Essentially the same result has been found by others using polystyrene standards on silica in *n*-hexane — dichloromethane mixtures (c_{crit} 59% DCM) [9], polystyrene standards on μ Bondagel in *n*-heptane — THF mixtures (c_{crit} 40% THF) [10], or polystyrene standards on silica in decalin — THF mixtures (c_{crit} 23% THF) [11].

1.2.3 Influence of Temperature

At the critical value of interaction energy, the enthalpy term just compensates for the entropy contribution, and samples of any molar mass are equally retained. The balance is limited to a certain temperature. Increasing temperature yields prolonged retention and vice versa (see Fig. 2). This might seem unusual at a first glance, but can be under-

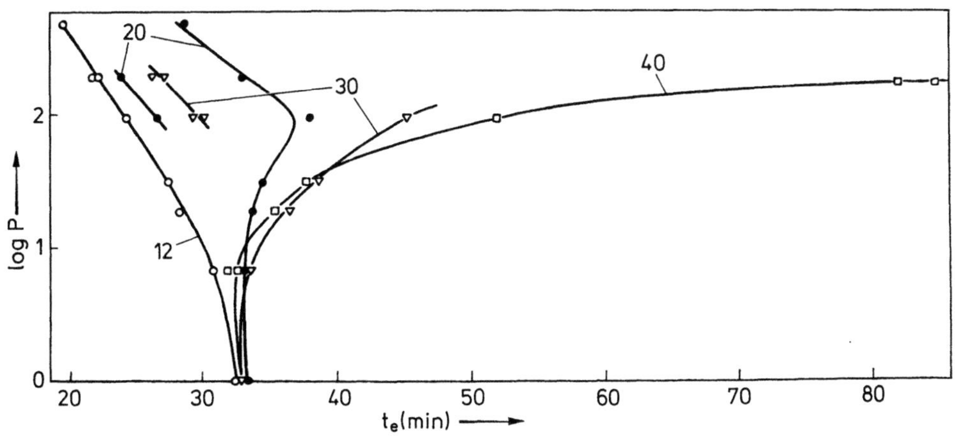

Fig. 2. Influence of temperature on the elution of polystyrene standards from a silica column. Eluent: 5.5% chloroform in tetrachloromethane; column as in Fig. 1. P — degree of polymerization. (From Ref. [7] with permission)

stood by taking into account the entropy balance of the whole system: the entropy increase caused by displacing solvent molecules from the surface is greater than the entropy loss by adsorption of a corresponding number of repeat units already interlinked in a chain polymer. In addition to this, increasing temperature conformationally favors polymer adsorption. If polymer retention in non-exclusion chromatography decreases with increasing temperature, the underlying retention mechanism is certainly not adsorption of the polymeric solute.

1.2.4 Gradient Requirements

Strongly adsorbed, flexible macromolecules will be generally not desorbed on lowering the concentration of the solute. This apparent irreversibility is the consequence of the interlinkage of the segments which in themselves are reversibly adsorbed. The removal of adsorbed macromolecules requires displacement by a stronger component, thus, gradient elution is typical of polymer chromatography in the non-exclusion mode. The variety of chromatographic mechanisms can be approximately classified by the type of stationary phase and the gradient necessary for a certain separation [12].

Gradient elution sets certain requirements with respect to the detector. Its signal must not be disturbed by the changing eluent composition and should be highly sensitive to the eluite. UV detectors are suitable for absorbing polymers in transparent fluids. Polymers with styrene units can be monitored at rather long wavelengths, e.g., 254 nm. Proteins and polypeptides in general require detection at 206–214 nm where peptide bonds absorb. Infrared detection offers the advantage of selective measurements of different constituents in copolymers, but has only limited sensitivity and requires solvents with an appropriate window. High sensitivity can be obtained through fluorescence detectors. The moving wire detector, which had been commercially available in the seventies, and the light scattering detector are suitable for any nonvolatile substance and compatible with any volatile eluent. In this respect they are universal for gradient chromatography of polymers.

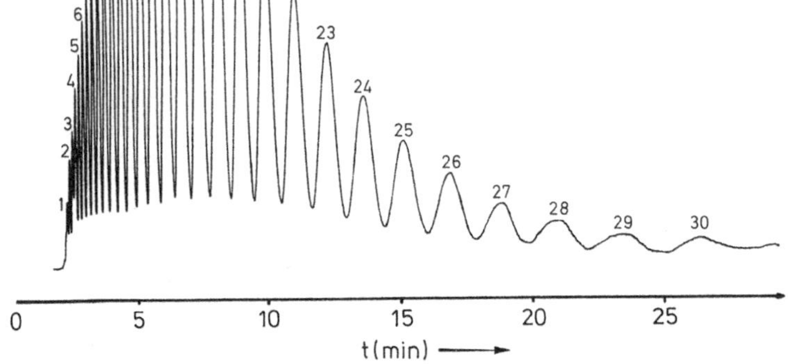

Fig. 3. Isocratic separation of oligomers from a polystyrene standard (M = 2100 g/mol) on a silica column (250 × 4 mm; d_O = 6 nm; d_P = 5 μm). Injection: m_0 = 0.3 mg in 10 μl; eluent: *n*-pentane — tetrahydrofuran (87:13, v/v); flow rate: 1 ml/min; UV detection at 254 nm. (From Ref. [14] with permission)

1.2.5 Column Packings

Gradient elution requires packing materials that can withstand the accompanied changes in solvent composition.

Another requirement is the accessibility of the active surface. A pore size of 30 or 50 nm is adequate in most cases of interactive polymer chromatography. The pore diameter must not be too large because the internal surface of the packing decreases with increasing pore diameter.

As already mentioned, uncontrolled superposition of size exclusion and interactive retention must be avoided since the order of separation in both types of chromatography is different. This is of particular importance in copolymer fractionation where macromolecules are to be separated which differ in size and chemical nature. Because of the strong influence of the latter on adsorption phenomena the results would be illusive if part of the sample was separated by exclusion and another by adsorption. Large-pore packings provide a chance to avoid interference as, in this case, all the

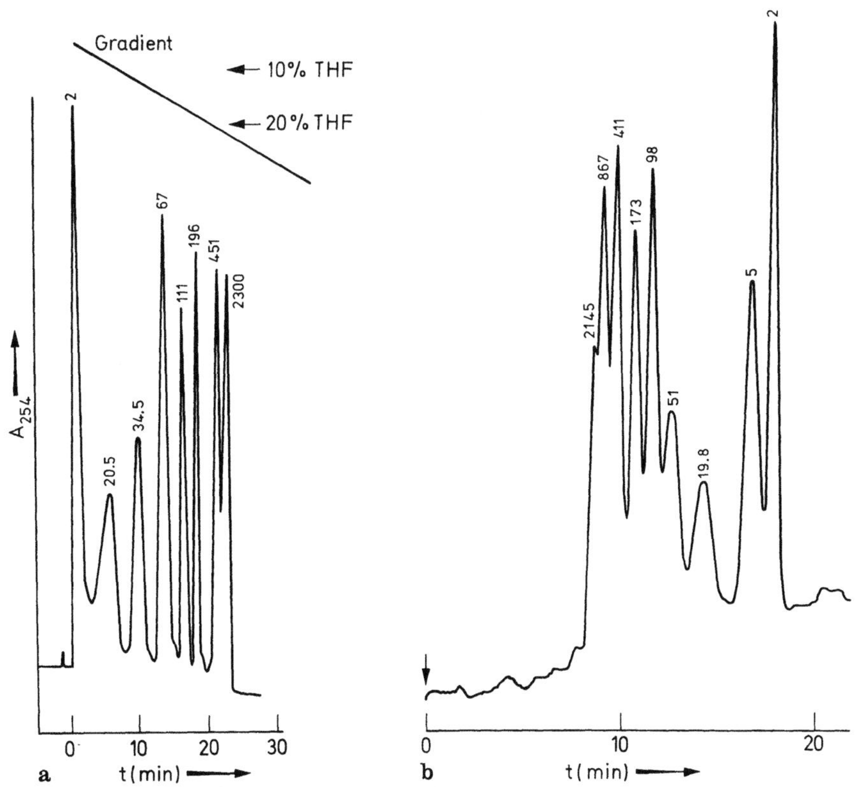

Fig. 4a. Adsorption HPLC of polystyrene standards. Column: silica (250 × 2.6 mm; d_O = 400 nm; d_P = 10 μm); gradient: decalin/tetrahydrofuran; flow rate: 0.5 ml/min; UV detection at 254 nm. (From Ref. [11] with permission). **b.** SEC of polystyrene standards. Column: silica (three columns in series, each 250 × 4 mm; nominal pore size 100, 500, and 1000 Å, respectively). In both chromatograms the figures at the peaks indicate the molar mass (kg/mol) of the respective polystyrene standard. (From Ref. [14] with permission)

molecules of the solute can penetrate. Another possibility is given through the use of packings with small pores inacessible to all of the large solute molecules.

1.2.6 Adsorption Versus Phase-Separation Mechanism

From profound experiences with thin-layer chromatography of polymers, it was concluded [13] that an adsorption "mechanism leads to separation largely according to the chemical composition of sample polymers" and that "TLC separation based on the phase-separation mechanism causes fractionation primarily by molecular weight differences." Unfortunately, this statement has been used by others sometimes in an absolute sense, i.e., omitting the qualifying adverbs "largely" and "primarily". Even worse, it has been used in such a manner as if "reversed-phase gradient elution" were synonymous with what was originally (and correctly) called "phase-separation mechanism".

In order to avoid further confusion, a few comments seem to be worthwhile:

1. There are numerous examples for isocratic oligomer separations by molar mass on polar adsorbents, i.e., under the conditions of adsorption chromatography. One of the first examples reported [14] is shown in Fig. 3.
2. Through gradient AC, even high-molecular polystyrenes have been separated by molar mass on polar packings [11] (see Fig. 4a). Figure 4b shows a similar separation by size-exclusion chromatography.

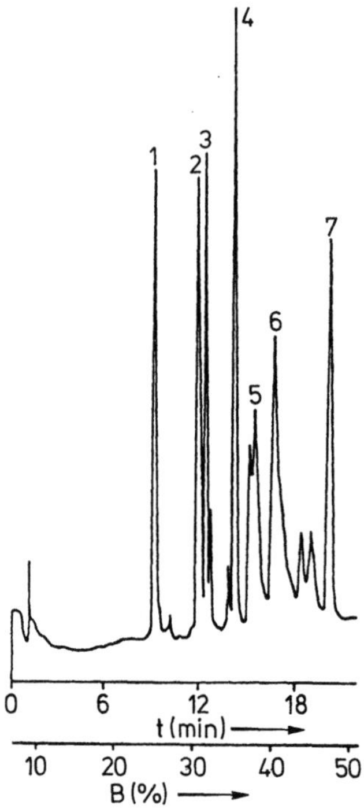

Fig. 5. Gradient elution of standard proteins on a Zorbax PSM C 18 column. Gradient: 0.01 *M* phosphate buffer pH 2/acetonitrile; samples: 1 — ribonuclease A, 2 — insulin, 3 — cytochrome C, 4 — lysozyme, 5 — transferrin, 6 — serum albumin, 7 — ovalbumin. (From Ref. [15] with permission)

3. Separation by composition has been performed not only on silica but also on non-polar carbonaceous materials and on bonded-phase packings of moderate polarity. The reversed-phase separation of biopolymers, especially of proteins, has gained great importance and is well-known to chromatographers and life scientists. What else should protein mapping be if not separation by composition? An example [15] of this important technique is given in Fig. 5.
4. Minimized interaction between eluite molecules and column packing is operative in hydrophobic interaction chromatography where biological solutes are so mildly bound to a stationary phase that their delicate native tertiary structure is not spoiled. This separation of extreme mildness is also a separation of chemically differing species, i.e., separation by composition.

Thus, the above-mentioned statement must not be generalized.

2 Investigation of Copolymers by Dual-Detection Size-Exclusion Chromatography or Thin-Layer Chromatography

Extensive work on copolymer composition distribution has been carried out using SEC apparatus equipped with an additional detector sensitive to only one kind of constituting unit. The advantage of this approach is the avoidance of extra experimental work. Its problems are connected with the evaluation of data [16]. The method will not be dealt with within these pages since it is a broad field in itself and has been reviewed recently [17–20].

Credit must also be given to TLC of copolymers which has contributed much to the present knowledge of the chromatographic behavior of polymers. Nevertheless, this method must be omitted as well; for comprehensive review see, e.g., the contributions of Inagaki [13,21] or Belenkii and Gankina [22].

3 Separation by Column Adsorption Chromatography

Porous silica is the most widely used sorbent in HPLC and has many advantages [23]. Its surface silanols are capable of adsorbing solutes according to their structure and polarity. In solution, the molecules of the solute and solvent compete for the active sites on the silica surface. The adsorption energy of solvents can be characterized by their ε^0 values [24]. The figures are usually tabulated for alumina adsorbents. For silica as an adsorbent, a plain proportionality is assumed [25]:

$$\varepsilon^0\,(SiO_2) = 0.77\varepsilon^0\,(Al_2O_3) \tag{10}$$

The following plain example might demonstrate the usefulness of the ε^0 data. In benzene ($\varepsilon^0 = 0.32$) polystyrene samples are eluted from a silica column, whereas polymethylmethacrylates and its copolymers are not. In THF (0.57) even PMMA homopolymers leave the column. Hence, THF is strong enough to prevent PMMA from adsorption. In chloroform (0.40) random copolymers with no more than 50%

MMA units are eluted but those richer in MMA remain on the silica. Under well-defined conditions, the influence of composition overcomes the influence of molar mass; but retention precisely independent of molar mass would be inconsistent with Martin's rule.

3.1 End-Group Analysis

Pioneering work was performed with polystyrene grafted onto cellulose. By hydrolysis of the cellulose backbone, polystyrene chains were obtained having glucose endgroups. The hydroxyls of the latter were strongly adsorbed in TLC on silica. Thus, the grafted chains could be separated from polystyrene by-products [26].

Polystyrene with carboxylic endgroups was investigated by adsorption chromatography on silica [27]. The endgroups were formed by the reaction of living polystyrene-Li with carbon dioxide. The goal of the preparative work was synthesis of polystyrene with one terminating carboxylic group per chain. The yield was greater than 90% when PS-Li and solid CO_2, or the Grignard compound polystyrylmagnesium bromide and CO_2 gas, were used. The analytical significance of the work is the separation and identification of by-products as, e.g., dipolystyryl ketone and tripolystyrylmethanol or compounds formed with oxygen. In TLC on silica the main product, polystyryl carboxylic acid, reached only low R_f values. The eluent was light petroleum — dichloromethane (30:70). The carboxylic acid from a 18,000 g/mol precursor reached a higher R_f than the corresponding product from a 2500 g/mol precursor. This might give the impression of a SEC separation, but, since the distyryl ketone and the tristyrylmethanol from the 2500 precursor reached higher R_f values than the 18,000 carboxylic acid, it is better understood as the consequence of different amounts of chromatographically mobile styrene units tugging at one anchoring group.

The 18,000 carboxylic acid has one polar group per 180 styrene units. The polar group is strong enough to determine the chromatographic mobility of the large molecule. This is just one example for the ability of AC to detect tiny differences in polymer structure.

The isolated 2500 polystyryl carboxylic acid appeared in two distinct peaks when investigated by HPLC on a silica column (150×4.1 mm; d_P = 5 μm) through a gradient *n*-hexane/diisopropyl ether saturated with water, 0–50% B within 15 min. The IR absorption of the material under both peaks was identical with that of polystyryl carboxylic acid. The authors concluded that the peak are due to diastereomers,

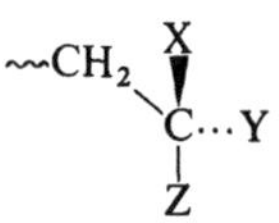

Fig. 6. Structure of the terminating group of a polymer chain

the structure shown in Fig. 6 having Z = C_6H_5, and X = COOH, Y = H or X = H, Y = COOH. At M = 2500 g/mol, about 50 C atoms form the backbone of the chain.

The ability of separating species differing in configuration at only one in 50 C atoms is a fine demonstration of the separation power of adsorption chromatography.

Oligomers from methyl methacrylate with OH endgroups were reasonably retained on a silica column (250 × 3.2 mm; $d_P = 5$ µm), but the corresponding acetal-terminated species eluted with the solvent [28]. In both cases, the eluent was ethyl acetate.

Methyl-capped oligomers could be separated into individual oligomers through a gradient butyl chloride/acetonitrile, 1–20% B in 20 min (see Fig. 7a). The same held true for the hydrogen-capped products which, interestingly, showed double peaks for each oligomer (see Fig. 7b). The splitting presumably indicates separation of diastereomers (cf. Fig. 6) with X = CH_3 and Y = H or X = H and Y = CH_3. The sample in Fig. 7a had X = Y = CH_3.

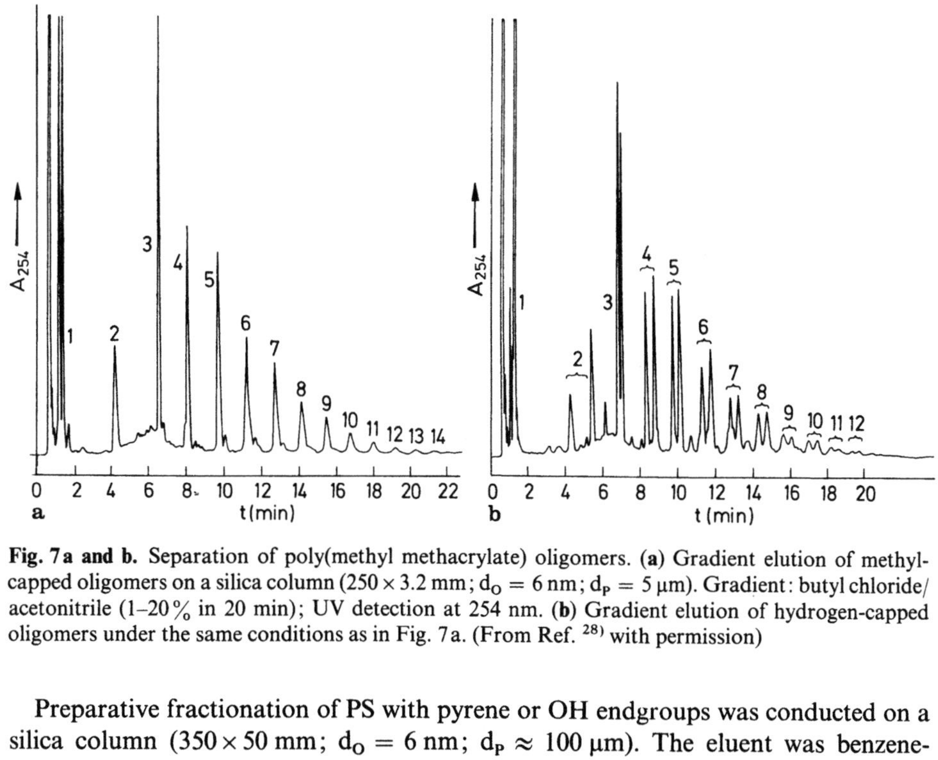

Fig. 7a and b. Separation of poly(methyl methacrylate) oligomers. (**a**) Gradient elution of methyl-capped oligomers on a silica column (250 × 3.2 mm; $d_O = 6$ nm; $d_P = 5$ µm). Gradient: butyl chloride/acetonitrile (1–20% in 20 min); UV detection at 254 nm. (**b**) Gradient elution of hydrogen-capped oligomers under the same conditions as in Fig. 7a. (From Ref. [28] with permission)

Preparative fractionation of PS with pyrene or OH endgroups was conducted on a silica column (350 × 50 mm; $d_O = 6$ nm; $d_P \approx 100$ µm). The eluent was benzene-cyclohexane (70:30) or chloroform with stepwise increased methanol content (0.01 to 2%). The fractions obtained formed a series decreasing in molar mass. This might have been due to size exclusion although the column possessed adsorption activity.

Unfortunately, the latter was unsufficiently controlled so that the technique remained "more of an art than a science" and about 20% of the sample were trapped in the column, i.e., the chromatographic recovery was 80%. In contrast to the elution order, the trapped material had the highest molar mass [29].

Polystyrene chains terminated with spiropyran groups were investigated on a silica column (250 × 4.6 mm; $d_O = 6$ nm; $d_P = 5$ µm). The living polymerization of the sample polymers was initiated by *n*-butyllithium and the termination reaction executed in the dark. Side reactions were proton termination and the coupling of PS precursor chains ($\overline{M}_n = 57{,}900$ g/mol). The spiropyran endgroups absorb at 350 nm and could thereby be monitored without interference from styrene units. Although the endgroup to repeat unit ratio was only 1:560, the labeled polymer was longer retained on the column than the hydrogenium-capped polystyrene. The gradient was *n*-hexane/dichloromethane increasing by 0.43%/ml from 62% DCM. Whereas this separation was not influenced by molar mass, a similar experiment using a column packed with silica of 50 nm pore size yielded retention of the unlabeled polystyrene increasing with molar mass. The gradient was also *n*-Hx/DCM with the steepness of 0.2%/ml, starting from 58% DCM [30].

3.2 Adsorption Chromatography of Copolymers

Mark and Saito [31] attempted fractionating polymers by means of chromatography as early as 1936. They filtered solutions of cellulose acetate in acetone through a column with a charcoal-like adsorbent made from blood. The eluate contained the fraction of highest molar mass; the rest of the sample was trapped in the column. It could be extracted by dioxane from separated portions of the packing. The largest molecules had travelled farthest. This was in contrast to expectation from Traube's rule and indicated size exclusion.

Beginning in the late forties, copolymers were fractionated by adsorption chromatography: poly(butadiene-*co*-styrene) [32–34], poly(butadiene-*co*-acrylonitrile) [32], poly-(styrene-*g*-vinyl acetate) [35], poly(styrene-*b*-ethylene oxide) [36] and poly(styrene-*co*-acrylonitrile) [37]. HPLC adsorption chromatography was first applied to copolymer analysis by Teramachi et al. in 1979 [38].

Porous silica is most widely used as adsorbent, but bonded phase materials with polar groups or crosslinked acrylonitrile [39] have also been tested. Silica requires painstaking control of activity. In the separation of poly(styrene-*co*-methyl methacrylate) samples with dichloroethane—chloroform mixtures, clearer results were obtained with a silica column previously rinsed with methanol [40]. Continuously decreasing activity of silica columns was observed in the elution of poly(styrene-*co*-methyl acrylate) with CCl_4-methyl acetate mixtures [38].

Gradient elution is much more problematic with silica than with bonded phase materials. Strongly adsorbed eluent components may be removed only with difficulties and may give rise to ghost peaks. Controlled water content of the eluents is helpful for obtaining a definite activity with reproducible retention values and, at the same time, is a powerful remedy against peak tailing [41]. The water is fixed to the stronger adsorption sites making the adsorbent more homogeneous. Only with water-saturated di-isopropyl ether as the eluent component B, separation of diastereomeric forms of oligostyrenes with carboxyl endgroups could be achieved [27], *vide supra*.

3.2.1 Poly(styrene-*co*-methyl acrylate)

Figure 8 shows the 259 nm UV record from the gradient elution of the mixture of three samples containing 46.6, 57.3, or 77.9 wt % methyl-acrylate units. This was the first report on copolymer separation by HPLC [38]. The chromatogram was obtained on a silica column (600 × 7.5 mm, d_O = 5 nm; d_P = 15 μm) with a CCl_4/methyl acetate gradient (7–35% in 35 min). In spite of the decaying activity of the column the composition distribution of the 57.3% sample could be evaluated.

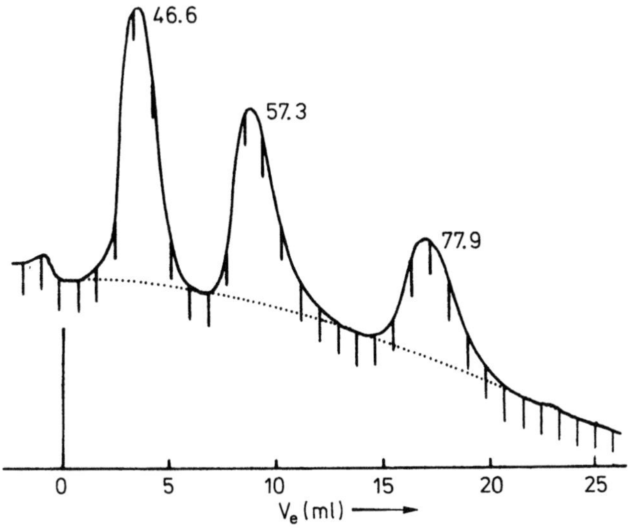

Fig. 8. Copolymer separation. Gradient elution of the mixture of three poly(styrene-*co*-methyl acrylate) samples on a silica column (600 × 7.5 mm; d_O = 5 nm; d_P = 15 µm). Gradient: tetrachloromethane/methyl acetate (7–35% B in 35 min); flow rate: 1 ml/min. The figures at the peaks indicate the composition of the respective copolymer in mol % methyl acrylate. Molar mass values: 46.6 — 261 kg/mol; 57.3 — 276; 77.9 — 302. (From Ref. [38] with permission)

3.2.2 Poly(styrene-*co*-methyl methacrylate)

Random copolymers from this system have been separated on a silica column (250 × 6 mm; d_O = 6 nm; d_P = 9 µm) with a dichloroethane/tetrahydrofuran gradient (exponential from 3 to 20%). The shape of the latter was recorded by means of a differential refractometer, whereas the elution of the sample was monitored by UV

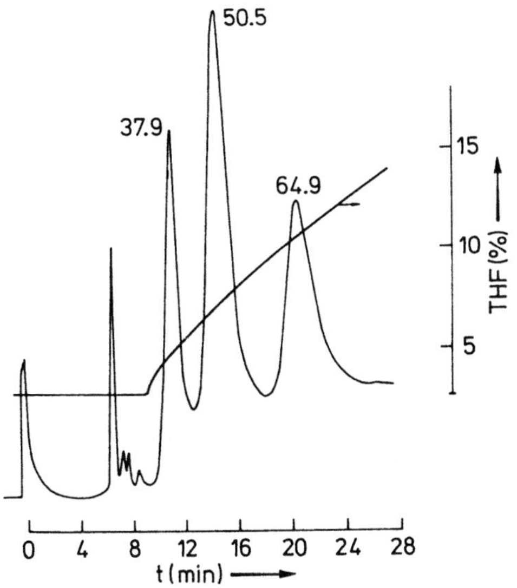

Fig. 9. Copolymer separation. Gradient elution of the mixture of three poly(styrene-*co*-methyl methacrylate) samples on a silica column (250 × 6 mm; d_O = 5 nm; d_P = 9 µm). Gradient: 1,2-dichloroethane/tetrahydrofuran (3–20% B, exponentially). Methyl methacrylate content (mol %) indicated. Molar mass values: 37.9 — 150 kg/mol; 50.5 — 185; 64.9 — 235. (From Ref. [42] with permission)

absorption. The mixture of three copolymers with 37.9, 50.5, or 64.9 mol % methyl methacrylate units and 150,000, 185,000, or 235,000 g/mol molar mass, respectively, could be separated (Fig. 9).

Before each analysis, the column was flushed with at least 10 column volumes tetrahydrofuran. Perfect reproducibility was observed during a period of several months [42]. Both dichloroethane and tetrahydrofuran are solvents for the copolymers investigated, thus, the retention and separation obtained was due to adsorption.

The mixture of seven random SMMA copolymers graded in methyl methacrylate content was separated on a silica column (150 × 4.6 mm; d_O = 6 nm; d_P = 5 μm) with a gradient of *iso*-octane/tetrahydrofuran + 10% methanol [43] (Fig. 10).

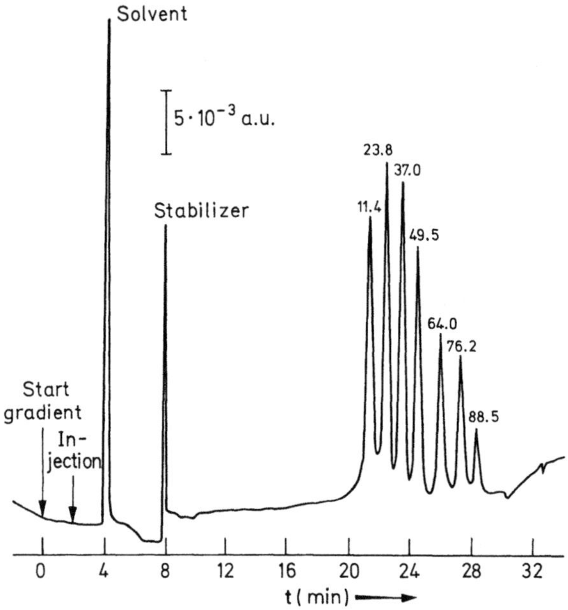

Fig. 10. Copolymer separation. Gradient elution of the mixture of seven poly(styrene-*co*-methyl methacrylate) samples on a silica column (150 × 4.6 mm; d_O = 6 nm; d_P = 5 μm). Gradient: *iso*-octane/(tetrahydrofuran + 10% methanol), 10% B (0 min), 50% (8 min), 80% (10 min), 100% (11 min); flow rate: 1 ml/min, reduced to 0.3 ml/min after 9.9 min. Methyl methacrylate content (wt %) indicated. Molar mass values: 11.4 — 160 kg/mol; 23.8 — 250; 37.0 — 150; 49.5 — 185; 64.0 — 235; 76.2 — 220; 88.5 — 220. Column temperature: 50 °C. (By courtesy of Elsevier Science Publ. [43])

Equally elegant chromatograms were obtained using cyclohexane/tetrahydrofuran [44] or *n*-hexane/tetrahydrofuran gradients. *Iso*-octane, cyclohexane, and *n*-hexane are nonsolvents for SMMA copolymers. Gradients with a high initial concentration of these liquids can cause precipitation of the injected sample. The higher the methyl methacrylate content of a copolymer is, the more tetrahydrofuran in mixtures with alkane hydrocarbons is needed to keep the copolymer in solution. This can be inferred from turbidimetric titrations. The more methyl methacrylate units a SMMA sample contains, the more THF is needed also for eluting it. Plotting the tetrahydrofuran concentrations from solubility investigations and chromatographic result vs. copolymer composition gives approximately parallel lines (Fig. 11). Obviously, solubility effects contribute to the separation in Fig. 10, but the main reason is believed to be adsorption. This follows from the higher amount of tetrahydrofuran in comparison with the solubility curve and the essential contribution of the polar packing material for this separation.

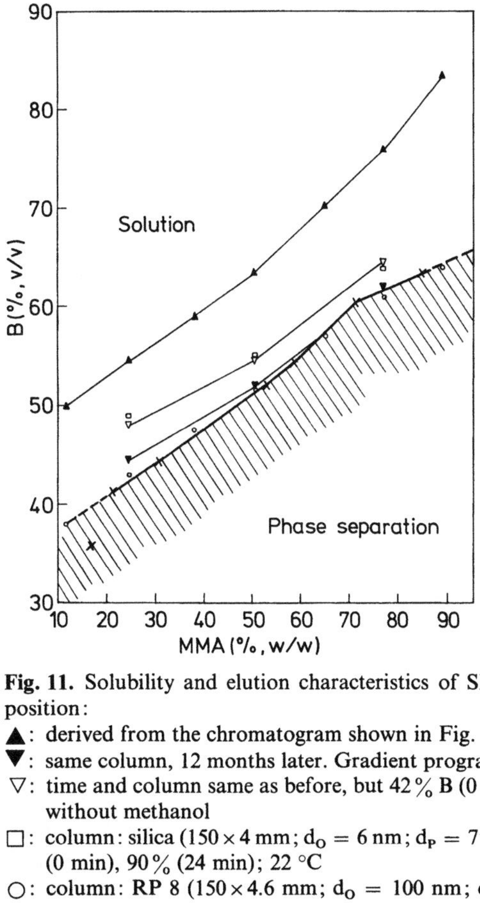

Fig. 11. Solubility and elution characteristics of SMMA copolymers. Eluent composition at peak position:

▲: derived from the chromatogram shown in Fig. 10

▼: same column, 12 months later. Gradient program: 40% B (0 min), 42% (20 min), 80% (25 min)

▽: time and column same as before, but 42% B (0 min), 44% (10 min), 100% (16 min) and B: THF without methanol

□: column: silica (150 × 4 mm; d_O = 6 nm; d_P = 7 μm). Gradient: *n*-hexane/THF; program: 30% B (0 min), 90% (24 min); 22 °C

○: column: RP 8 (150 × 4.6 mm; d_O = 100 nm; d_P = 10 μm). Gradient: *n*-hexane/(THF + 10% methanol); 10% B (0 min), separating range 33–80% B within 10 min.

Flow rate: 1 ml/min (note specifications mentioned for the chromatogram shown in Fig. 10).

×: Solvent composition at precipitation points from turbidimetric titration at 20 °C with *n*-hexane precipitant and THF solvent. These points determine the boundary of the solubility region

With a silica column (50 × 4.6; d_O = 3 nm; d_P = 5 μm) and the eluent system dichloroethane — chloroform, only stepwise elution of SMMA copolymers was possible. With pure dichloroethane as an eluent, the mixture from three copolymers containing 14, 27, or 34% MMA units was trapped in the column. (The molar mass of the first both was 77,000 or 79,000 g/mol, respectively). An eluent containing 30% chloroform delivered a sample with 14% MMA, by 60% chloroform the 27% MMA specimen was eluted, and pure chloroform brought the rest (34% MMA) out of the column. This elution behavior was found if the column had been flushed with methanol. Without this pretreatment, the 34% sample remained in the column even in pure chloroform that eluted just the 27% sample. The 14% specimen was eluted in two portions by application of 30 and 60% chloroform [40].

SMMA copolymers in the composition range 19–76 mol % MMA could be separated on a 500 × 7.5 mm column packed with crosslinked copolymer from acrylonitrile and ethylene dimethacrylate. A gradient *n*-hexane/chloroform (exponentially from 35 to 85%) allowed five specimens with 19, 44, 62, and 76 mol % methyl methacrylate to be separated [45].

3.2.3 Poly(styrene-*co*-butadiene)

The mixture of six SB samples with 11, 21, 33, 41, 58, or 82 mol % styrene was separated on a column packed with a copolymer from 67% (v/v) acrylonitrile and 33% ethylene dimethacrylat (500 × 7.5 mm; d_P = 2.5–5 µm). The $\overline{M}_n$ values of the samples were 46, 63, 69, 112, 180, or 207 kg/mol, respectively.

Separation was obtained through *n*-hexane/chloroform gradient (exponentially from 10 to 45%). Since 15% chloroform sufficed for disolving samples with no more than 33% styrene, the authors concluded that adsorption was responsible for the separation obtained [39].

3.3 Fine Resolution of Oligomers

About 40 individual oligomers could be isolated from a polystyrene calibration standard with a nominal molar mass of 2100 g/mol. This was performed isocratically on a silica column (250 × 4 mm; d_O = 6 nm; d_P = 5 µm) through a *n*-pentane-tetrahydrofuran mixture (87:13) [14] (see Fig. 3), as well as by gradient elution with *n*-hexane/tetrahydrofuran or *n*-hexane/ethyl acetate on a silica column (250 × 4.6 mm; d_O = 6 nm, d_P = 5 µm). With a weaker eluent B, e.g., dichloromethane, the fine structure of the first peaks could be revealed which can be attributed to stereoisomers. In a physical sense, these are copolymers uniform in constitution but differing in configuration. With two silica columns in series (each 250 × 4.6 mm; d_O = 6 nm; d_P = 5 µm) and a gradient *n*-hexane/dichloromethane, 9 of 10 possible hexamers and 11 of 20 possible heptamers could be separated [46] (Fig. 12). The sample was a polystyrene standard with 800 g/mol nominal molar mass, the gradient program an increase of B concentration by 0.1%/min from 11% at t = 0, the flow rate 1 ml/min.

Isocratic elution with *n*-hexane — dichloromethane (78:22) also disclosed the existence of stereoisomers [47].

Both the highly active adsorbent and the moderate eluent seem to be necessary for the separation of oligostyrenes into stereoisomers. Elution of another sample on a nitrile bonded-phase column with *iso*-octane/dichloromethane could not reveal stereoisomers but only well-separated oligomers [48].

Oligomers prepared by radical polymerization of 2-vinylpyridine could be separated into tactic isomers on a silica column (250 × 4.6; d_O = 10 nm; d_P = 10 µm) with a gradient pentane/methanol (concave 10 to 40% in 30 min). The experiments were conducted under supercritical conditions at 250 °C and 66 bar using a modified commercial apparatus [49]. Unfortunately, the silica was not stable under these conditions and melted.

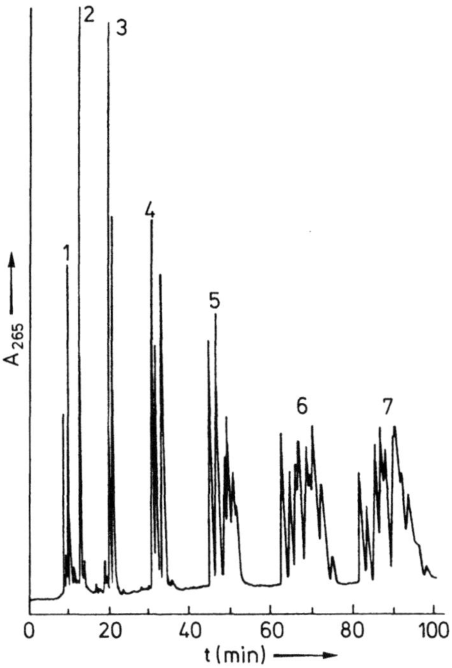

Fig. 12. Isomer separation. Gradient elution of 50 µg polystyrene 800 on a tandem of two silica column (2 × 250 × 4.6 mm; d_O = 6 nm; d_P = 5 µm). Gradient: *n*-hexane/dichloromethane, 11–21% B in 100 min; flow rate: 1 ml/min; UV detection at 265 nm, 0.05 AUFS. (From Ref. [46] with permission)

4 Separation by Reversed-Phase Chromatography

The term reversed-phase chromatography was coined by Howard and Martin in 1950 [50] when liquid chromatography was usually performed on a polar stationary phase with a less polar eluent. If a gradient was used, it would have been one of increasing polarity. This procedure was normal at that time. The term reversed-phase chromatography should indicate separations using a stationary phase of moderate polarity and a more polar eluent and, in the case of gradient elution, the decreasing polarity of the mobile phase. Currently about 80% of all HPLC separations are conducted in RPC mode, but nevertheless the chromatography on polar stationary phases is referred to as 'normal'.

RP-HPLC is mostly performed on hydrocarbonaceous layers chemically bound to porous silica. Its silanol groups can react with a variety of organosilyl compounds, e.g., with $(H_3CO)_2Si(CH_3)R$.

With R = $-(CH_2)_{17}CH_3$, the final product is octadecyl silica stationary phase (ODS, C18, RP 18). Steric hindrances or other reasons leave some fraction of the silanol groups unreacted and additional ones are produced by hydrolysis of superfluous $SiOCH_3$ groups. The silanol groups from both sources can be subject to capping reactions with N,O-bis(trimethylsilyl) acetamide, $Cl-Si(CH_3)_3$, or $(CH_3)_3Si-NH-$

$Si(CH_3)_3$. But even endcapped bonded phases still have free silanols. The structure of bonded-phase materials has been studied by various methods including gas chromatography of the organic ligands after digestion of the supporting silica [51, 52] and solid state ^{13}C and ^{29}Si cross-polarization and magic-angle spinning NMR [53].

The unreacted silanols are hidden in a thicket of carbonaceous residues whose bulkiness and sweep determine the accessibility of the silanols. The extension and intensity of the sweep will be influenced by mobile phase composition and temperature.

The main reason for chromatographic retention on reversed phases is solvophobic interaction, but under certain conditions silanophilic interaction and additional effects contribute to, or even govern, the chromatographic process. A concise review of physicochemical phenomena contributing to retention in RPC has been recently given by Horváth [54].

In comparison with the popularity of RPC in low-molecular HPLC it is quite unexpected that this technique has been scarcely used for the investigation of synthetic polymers so far. Before discussing this, let us first mention the importance of RPC in the separation of proteins and other biopolymers.

4.1 Separation of Proteins and Polypeptides

The isolation and characterization of biologically active compounds is possibly the most challenging task in HPLC and an important contribution to progress in life science and biotechnology. In addition to this, the wealth of experience gathered in this field must be known by scientists considering non-exclusion HPLC of any polymer.

In terms of polymer science, proteins are copolymers from amino acids having the unique advantage of definite structure and size. In contrast to synthetic polymers, proteins are really chemically pure substances. Under physiological conditions, most of them even have definite conformation. Preserving this tertiary structure is often problematic. Unfolding is connected with denaturation. The sensitivity of the primary structur against degradation by chemical or biological attacks makes the research additionally difficult — but at the same time even more challenging.

4.1.1 Basic Decisions

The delicate structure of proteins makes it difficult to achieve all goals of a separation procedure simultaneously. At the moment, the best approach is to find the main goal of a separation and to set the chromatographic conditions according to it. Protein mapping for identification and analytical information, or the isolation of proteins for subsequent chemical analysis of the ratio and sequence of the amino acids can be conducted under conditions that violate the tertiary structure. Separations for this purpose can be performed in reversed-phase mode using organic solvents.

On the other hand, if the separated proteins must maintain the biological activity, e.g., in the purification of enzymes, the conditions of separating must be mild. Aqueous eluents of approximately physiologic pH value are required and only moderate interactions between the eluite and the stationary phase should occur. Hydrophobic interaction chromatography (HIC) or ion exchange chromatography (IEC) are appropriate.

4.1.2 Brief Description of HPLC Techniques for Protein Separation

Typical of *reversed-phase chromatography* is the use of a nonpolar column, e.g., C 18 bonded to silica, and a gradient water/organic modifier. Packing materials of 30–50 nm pore size are recommended and the carbon load of the stationary phase should better be moderate (2–4%, w/w) than maximum [15]. Endcapping is required. The modifier is an organic liquid miscible with water. Propanol, acetonitrile, tetrahydrofuran, or methanol are used, decreasing in eluotropic strength from propanol to methanol. In general, best resolution is obtained with modifier concentrations between 15 and 40%. This condition implies the selection of a modifier with an appropriate elution strength.

Acetonitril is often recommended because of its separating power (cf. Fig. 5). Adding a small amount of trifluoroacetic acid (TFA) to both gradient components A and B, is state of the art in protein reversed-phase chromatography. It decreases retention time, improves resolution, and increases the recovery of the sample proteins.

The use of RP columns, ion pairing additives, and a gradient ending with rather high concentration of organic modifier are, of course, only rough sketches of a detailed picture.

Hydrophobic interaction chromatography is carried out on porous hydrophylic materials supporting hydrophobic ligands at a rather low density (1–10% of the ligand density suitable for RPC). Alkylated agarose gels or the TSK Phenyl 5-PW columns, e.g., are commercially available. The packings of the latter have phenyl groups (about 0.1 mol/ml) linked by ether groups to TSK gel G 5000 PW which is a polymer-based hydrophylic material of large pore size and $d_P = 10\ \mu m$.

HIC columns are used with aqueous gradients of decreasing salt concentration, e.g., 2 *M* Na_2SO_4/water. To maintain pH control, the gradient components contain buffer, e. g., 0.01 *M* potassium phosphate.

The physical phenomena causing retention in HIC are basically the same as in solvophobic RPC. In HIC, the moderate binding power of the packing is complemented by the salting-out effect of the mobile phase. The mild conditions of HIC provide the best chance for recovering proteins of biological activity.

The retention caused by different salts in equal concentration is in accordance with the Hofmeister series. The higher the molar surface tension increment of a salt, the greater is its effect on protein retention [55]. Higher column temperatures produce prolonged retention in HIC, in contrast to RPC where retention decreases with increasing temperature [56].

In *ion exchange chromatography*, weak cation exchanger (e.g., with carboxylic acid groups) or anionic exchanger (e.g., with aminoethyl groups) are used as stationary phases. The buffered aqueous mobile phase is a gradient increasing in salt concentration, e.g., 0/0.2 *M* NaCl in 0.1 *M* Tris-HCl. Retention is based upon coulombic interaction and thus basically different from those in RPC or HIC. Under certain conditions, RPC separations may partly assume IEC behavior, and vice versa. For example, collagen chains could not be eluted from a wide-pore CN or C18 bonded-phase column by common RP gradients (water + TFA)/organic modifier. The sample remained on the column due to excessive ionic interactions between protein and support. The addition of ammonium bicarbonate (0.02 to 0.5 *M*) to the aqueous eluent overcame these difficulties. Best resolution was obtained through 0.05 *M* salt [57]. The counterpart to this RP + IE behavior is the observation that angiotensin peptides could be separated on a weak cationic exchange column by an ammonium formate gradient only after the superposition of an acetonitrile gradient. This indicates solvophobic interactions in addition to coulombic retention [58].

Obviously, mixed-mode retention can be advantageous in some difficult separation problems. Of course, this remark does not imply cancelling the well-established rules in HPLC.

4.1.3 Unfolding Under RPC Conditions

The tertiary structure of native proteins is stabilized through hydrophobic interactions in the interior of the three-dimensional structure. The strongly hydrophobic conditions of RPC are known to unfold this conformation. With some species, e.g., with lysozyme, the unfolding is reversible. Here, even RPC does not produce permanently deactivated species; but, in general, unfolding causes denaturation.

The elution pattern in IEC results from the charge distribution on the folded chain. Therefore, IEC was used for indication, whether the native structure of the protein had been affected by previous RPC or not. Ribonuclease was found to retain its native structure, whereas bovine serum albumin, horse radish peroxidase, and ovalbumin were much altered through RPC on a C 18 column with a gradient water/(ethanol-butanol 80:20) containing 0.012 *M* HCl in both eluent components [59].

In RPC on a C 4 column (100 × 4 mm; d_O = 50 nm; d_P = 10 μm) and a gradient water/(water + organic modifier), A and B with 0.01 *M* H_3PO_4, two distinct peaks were observed on injection of either papain (120 μg), lysozyme (50 μg), or α-chymotrypsinogen (120 μg) [5]. The first of both peaks was always due to the active form, the latter eluting one to the denaturated protein. The denaturation showed two steps, the first one occurring immediately upon contact between the protein and the stationary phase, and the other one, with a half-life of about 15 min, caused by the organic modifier. Figure 13 shows that at constant surface contact time, the denaturated (second) peak was smaller if some 1-propanol had already been in the column when the sample was injected. Obviously, hydrophobic interaction between 1-propanol and C 4-ligands can make the bonded layer softer, i.e., more hydrophilic towards a protein. Prolonging the surface contact by delaying the gradient start beyond the injection time caused additional denaturation (Fig. 14). Plotting the area of the native (first) peak vs the incubation time enabled the first-order rate constant k of the denaturation process to be estimated. k increased with the temperature and with the propanol content of the mobile phase during the incubation period. The latter influence is due to the well-known denaturating effect of organic solvents. It occurred in addition to the fast contact denaturation shown in Fig. 13.

Denaturation in solution is known to be faster with 1-propanol than with 2-propanol [60], but in contact with the C 4 RP column a higher rate was observed with 2-propanol. Acetonitrile produced the highest acceleration. Replacement of H_3PO_4 by 0.01 *M* trifluoroacetic acid did not alter the 2.2 pH, but enhanced by a factor of three the rate of denaturation at 22 °C. Thus, eluent systems containing TFA and acetonitrile are extraordinarily active in denaturation. The wide popularity of just these systems is evident if one considers eventual refolding. If the refolding rate as well as the rate of unfolding are low, two separate peaks will occur (cf. Figs. 13 and 14) and, if the half-life of refolding is of the same order as the chromatographic elu-

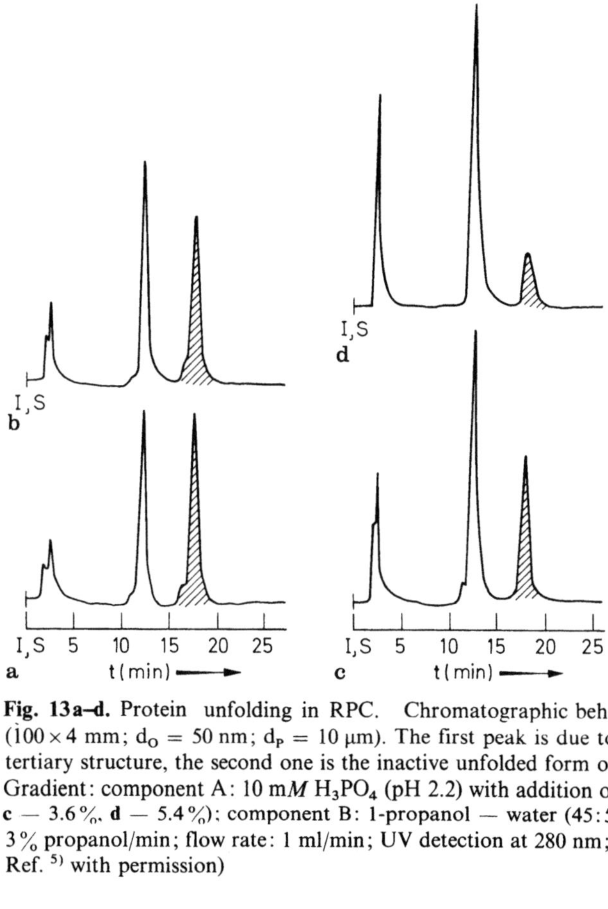

Fig. 13a–d. Protein unfolding in RPC. Chromatographic behavior of papain on a RP 4 column (100 × 4 mm; d_O = 50 nm; d_P = 10 μm). The first peak is due to the active protein with the intact tertiary structure, the second one is the inactive unfolded form of papain. Injection: 120 μg in 6 μl. Gradient: component A: 10 m*M* H_3PO_4 (pH 2.2) with addition of 1-propanol (**a** — 0%, **b** — 1.8%, **c** — 3.6%, **d** — 5.4%); component B: 1-propanol — water (45:55) with 10 m*M* H_3PO_4; program: 3% propanol/min; flow rate: 1 ml/min; UV detection at 280 nm; column temperature: 5 °C. (From Ref. [5]) with permission)

tion time, distorted peaks will result in a RPC system executing slow unfolding. Hence, driving the protein molecules in the fully unfolded state is preferred in order to obtain only the single sharp peak from the denaturated form.

The ion-pairing activity of trifluoroacetic acid renders this agent an almost inescapable concomitant in optimizing RPC separation. Interestingly, it does not always annihilate the biologic activity of the sample under investigation. This was observed in the RPC of human chorionic gonadotropin (hCG, a hormon of placental origin containing 30% carbohydrate, M = 65,000 g/mol) on a C 18 column (250 × 4.1 mm) using a gradient water/(ethanol — butanol 70:20). Both gradient components contained 10% ethyleneglycol dimethylether and a small amount of TFA. With 0.025% in A and 0.006% in B, a relative biologic potency of 0.20 was obtained. Separations in this eluent system, where the TFA content increased by a factor of 2 or 4, led to a biological potency of 0.45 or 0.54, respectively [61].

The reasons causing denaturation in RPC are obvious: low pH, low ionic strength, organic eluents, and, last but not least, the contact with the rigid and hydrophobic surface. The question how to suppress the unfolding is of practical importance. Low temperature and short contact time with the hydrocarbonaceous layer are recommended. Bonded phases with moderate hydrophobicity may be advantageous, e.g., phenyl-

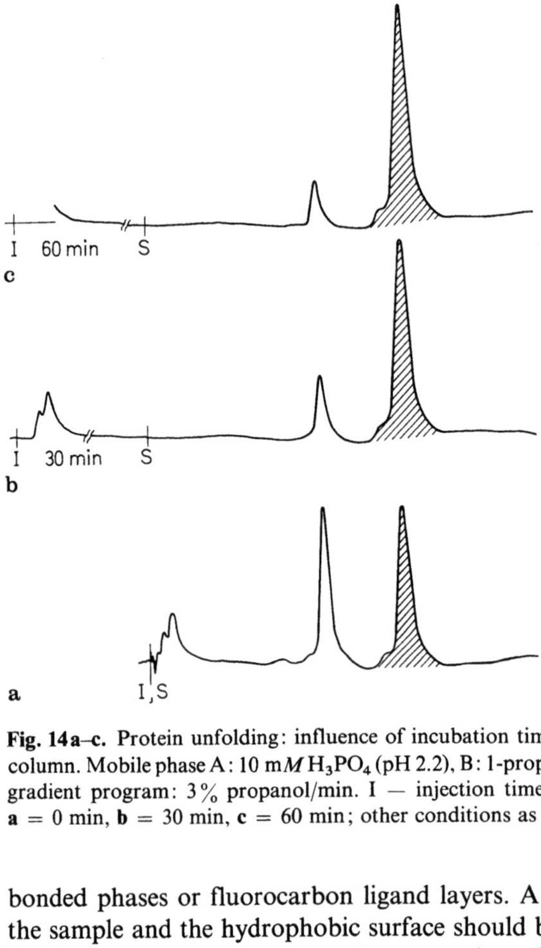

Fig. 14a–c. Protein unfolding: influence of incubation time. Gradient elution of papain on a RP 4 column. Mobile phase A: 10 m*M* H_3PO_4 (pH 2.2), B: 1-propanol — water (45:55) with 10 m*M* H_3PO_4; gradient program: 3% propanol/min. I — injection time, S — start of the gradient. Incubation: **a** = 0 min, **b** = 30 min, **c** = 60 min; other conditions as in Fig. 13. (From Ref. [5] with permission)

bonded phases or fluorocarbon ligand layers. A too drastic confrontation between the sample and the hydrophobic surface should be avoided. Thus, injection into the running gradient is recommended, or a starting mixture with some modifier [5]. Several authors suggested the addition of surfactants. The idea was supported by isocratic retention measurements of proteins on RP columns in KH_2PO_4 — 2-propanol mixtures as a function of the surface tension of the eluent [62]. The onset of elution is bound to a drop of surface tension. Chromatographic results with surfactants reported so far show either a rather stable coating of the RP packing leading to size exclusion separation or severe peak broadening and tailing [63].

Currently, the best way of preserving the native conformation under the conditions of elution chromatography is hydrophobic interaction chromatography with minimized adsorption forces. These can be adjusted by selecting the appropriate stationary phase, e.g., from a series of alkylated agarose gels with various alkyl chain lengths.

Then desorption can be obtained by very moderate influences, such as decreasing the ionic strength, changing the ionic quality, altering the pH towards the isoelectric point of the retained protein, lowering the temperature, adding polarity-reducing agents, mild detergents, or specific deformers (e.g., imidazolum citrate) [64].

4.1.4 Recovery

Another and even more basic problem is protein recovery. This problem is not serious in peptide chromatography. With [^{125}I] angiotensin II (M = 1170 g/mol) RPC was possible on a C 8 column (150 × 4.6 mm, d_P = 5 μm) with nearly full recovery at the picogram and femtogram level, but a brandnew column from the same supplier

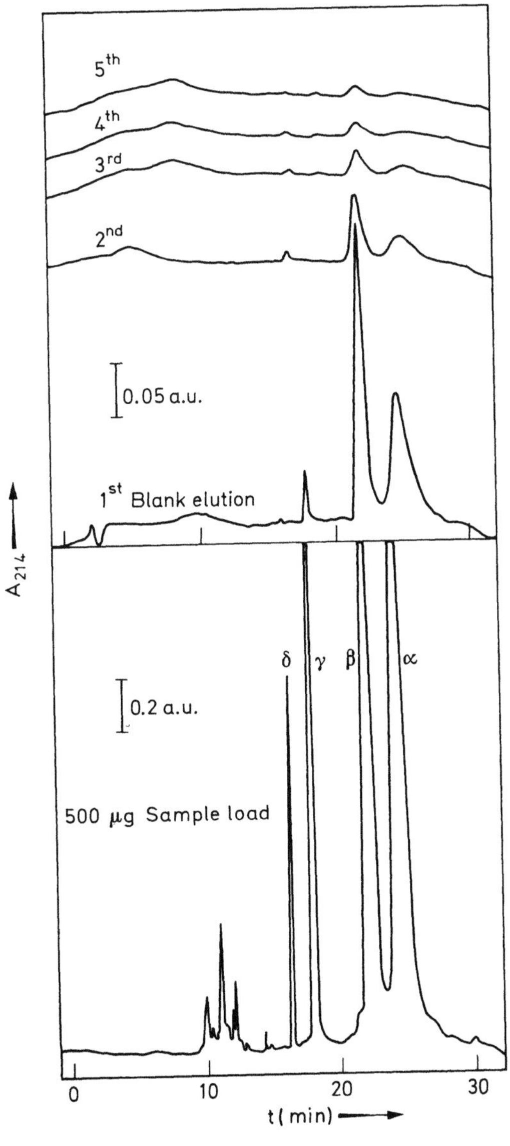

Fig. 15. Memory effect. Gradient elution of 500 μg phosphorylase kinase on a RP 18 column (250 × 4.6 mm; d_P = 5 μm). Mobile phase A: 0.1% TFA in water; B: 0.08% TFA in acetonitrile; gradient program: 0% B (0–8 min), 46% (9 min) 68% (24 min), 75% (33 min); flow rate: 1 ml/min. The lower chromatogram was obtained upon sample injection; the five blank gradients were performed immediately after the initial separation. Molar mass of the subunits: α — 132,000 daltons; β — 113,000: γ — 43,000; δ — 16,680. (From Ref. [66] with permission)

and of the same nominal quality gave tenfold increased retention and only 68% recovery [65].

With the increase of molar mass, recovery becomes a problem. The lower part of Fig. 15 shows the chromatogram of 0.5 mg phosphorylase kinase on a C 18 column (250 × 4.6 mm; d_P = 5 μm) with a gradient water/acetonitrile with 0.1% TFA in eluent A and 0.08% in B. The α, β, γ, δ subunits have molar mass values of 132, 113, 43, and 16.68 kg/mol, respectively, and are incorporated in equal molar amounts. The tetrameric holoenzyme $(\alpha\beta\gamma\delta)_4$ was estimated to have a molar mass of 1.2–1.3 × 10^6 g/mol. The upper part of Fig. 15 shows the UV record of five blank gradient cycles immediately following the initial separation. The larger subunits are desorbed more slowly than the lighter ones. The paper reports 76% recovery from that column on injection of 0.25 mg. The results of Fig. 15 suggest flushing the column with a strong eluent between each separation or performing a series of blank gradients in order to maintain resolution, recovery, and column life [66]. On a C 4 column (250 × 4.6 mm; d_P = 5 μm), analytical sample amounts (5–30 μg) were recovered at an average of 55% with an only minor influence of the sample load. Best results for micropreparative separations were obtained on a C 4 column (60 × 10 mm; d_P = 5 μm) with an average recovery of 76% for the total protein after injection of 0.422 mg. The recovery of the four subunits of phosphorylase kinase increased from α to δ, i.e., with decreasing molar mass (64, 71, 79, 89%, respectively). The authors recom-

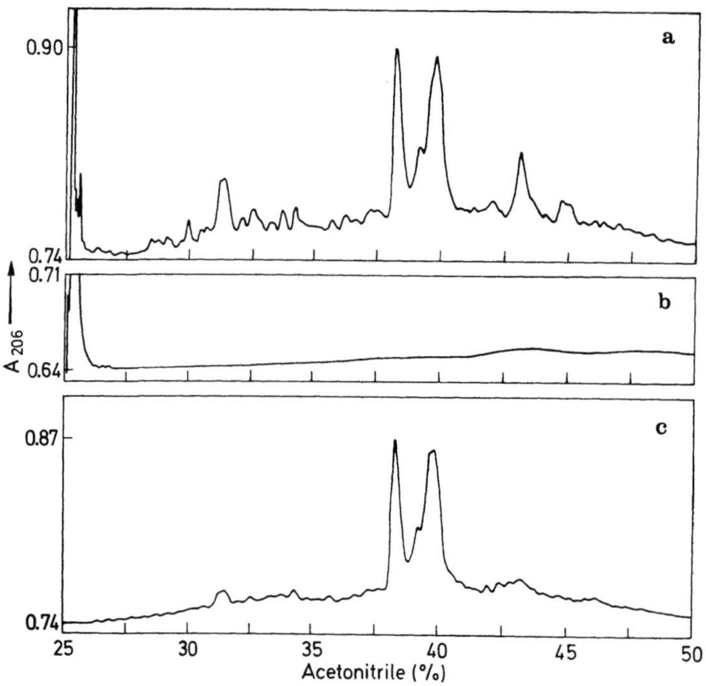

Fig. 16a–c. Influence of TFA. Gradient elution of histone proteins (m_0 = 467 μg) on a RP polystyrene-gel column (150 × 4.1 mm; d_P = 10 μm). Gradient: water/acetonitrile; gradient rate: 10% B/h; flow rate: 1 ml/min; UV detection at 206 nm. **a** = gradient with 0.1% TFA; **b** = gradient without TFA: the injected protein was not eluted; **c** = blank gradient with 0.1% TFA following run (**b**). (From Ref. [68] with permission)

mended limiting the sample load to 0.3 mg for a column with 4.6 mm internal diameter and to 0.5 mg for a 10 mm column [66].

Recovery of only 20% was observed in the separation of influenca virus components on a RP column (250 × 4.6 mm; d_O = 30 nm) with a water/acetonitrile gradient. The virus consists of three protein components with molar mass values ranging from 28 to 55 kg/mol. After several runs, the column back pressure increased from 60 to 90 bars. Overnight treatment with 0.1% sodium dodecylsulfate and 0.05% TFA helped to regain approximately the initial permeability [67].

Attempts to separate histone proteins by HPLC were faced with many difficulties caused by irreversible adsorption on reversed-phase packings. The histones belong, together with DNA, RNA, and non-histone proteins, to the chromatin of the cells and are known to function as structural components that organize the DNA into nucleosomes. The histones may also control several DNA activities. The problems in RPC separation of histones could be overcome if (1) silica-based RP materials had been endcapped and (2) trifluoroacetic acid (or a similarly ion-pairing agent) was added [68]. This investigation also shed light on the function of TFA in such separations. Even polystyrene gel, which is assumed to be free of polar groups, trapped the histone proteins without addition of TFA (see Fig. 16).

4.1.5 Sample Solubility

The limited solubility of membrane proteins and related polypeptides in aqueous mobile phases can also cause problems. These could be solved, e.g., by adding guanidine hydrochloride (6 *M*) or urea (8 *M*) to the portion of initial eluent used for sample preparation [69]. The urea was always eluted in the breakthrough volume of the column. Thus, the retained hydrophobic polypeptides might have been temporarily precipitated upon the column. Collagen chains, dissolved in 0.5 *M* acetic acid, were successfully separated by RP-HPLC through gradients of 0.1 *M* TFA/acetonitrile [70] or (0.05 *M* ammonium bicarbonate + TFA)/ tetrahydrofuran [57].

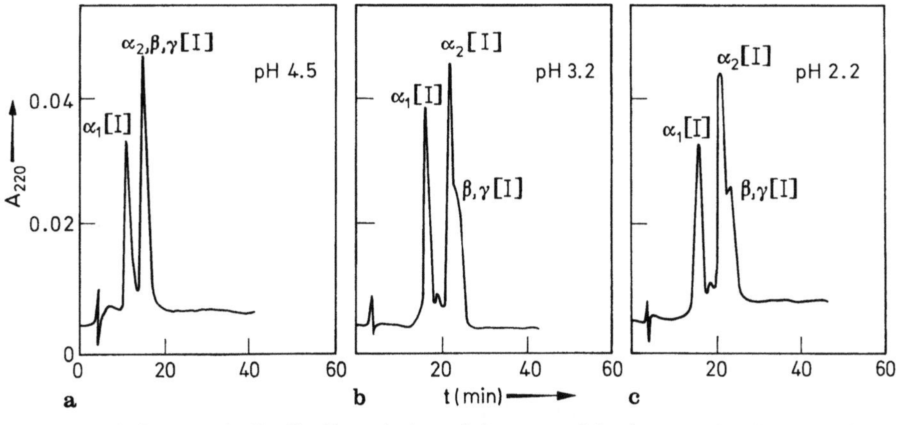

Fig. 17a–c. Influence of pH. Gradient elution of denaturated bovine type I collagen chains on a CN-bonded phase column (250 × 4.6 mm, d_O = 33 nm) at different pH. Mobile phase A = 0.05 *M* ammonium bicarbonate in water + TFA, B = THF; gradient program: from 20 to 30% B in 60 min; flow rate: 1 ml/min; column temperature: 35 °C. The pH was adjusted to the given value by addition of TFA. (From Ref. [57] with permission)

Poorly soluble virus proteins have been separated by RPC through gradients of formic acid/acetonitrile. Even under these harsh conditions about 100% recovery was found [71].

4.1.6 The pH Effect

The eluent is rather acidic in most protein separations. With silica-based packings, the reason for the selection of this condition is related to the surface silanols which are weakly acidic groups. Above pH 3.5–4, these groups are dissociated and ionic interactions will occur with basic sample compounds. These interactions can be reduced simply using acidic mobile phases that prevent ionization of the silanols. Another reason is the stability of the silica which is higher in acidic media [72].

An example for the influence of the pH was given in a recent investigation of collagen chains on a CN column (250 × 4.6 mm; d_O = 33 nm) through a water/tetrahydrofuran gradient. The aqueous medium contained 0.05 *M* NH_4HCO_3 and an amount of TFA necessary to adjust a certain pH value. The separation occurred by a mixed mode mechanism, including ionic or electrostatic interactions, because the collagen could be eluted only after the addition of the ammonium bicarbonate. Figure 17 shows the influence of the pH and the improved resolution in acidic media [57].

4.1.7 Influence of Column Length

In gradient elution of proteins it was often observed that the length of the column did not effect the resolution.

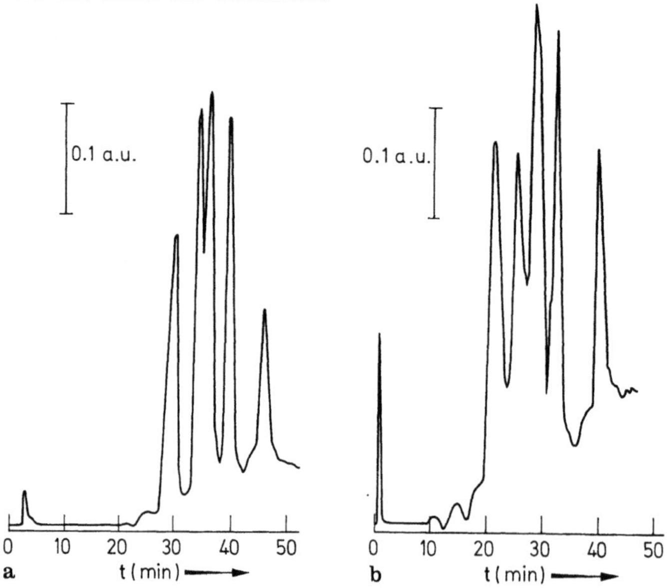

Fig. 18a and b. Influence of column length. Gradient elution of a protein mixture on RP 8 columns of different length (**a** = 45 × 4.6 mm, **b** = 6.3 × 4.6 mm; d_O = 30 nm). Gradient: water/2-propanol, both with addition of 0.1% TFA; gradient rate: 4%/ml; flow rate: 0.25 ml/min. The sample was a mixture of ribonuclease A, insulin, cytochrome C, bovine serum albumin, and ovalbumin (in the order of elution). These components were applied in the ratio 2:2:1:2:2 and totaled 50 μg. UV detection at 280 nm. (From Ref. [73] with permission)

The separation of protein test mixtures on very short C 8 columns yielded surprising results (Fig. 18). The resolution obtained with a 6.3 × 4.6 mm column was about 2.25 or 2.22 times better than with a 45 × 4.6 mm column when the flow rate was low (0.25 or 0.5 ml/min, respectively) and the volumetric gradient rate large (16% B/ml). The gradient components were A: water and B: 2-propanol, both with 0.1% trifluoroacetic acid (TFA) [73]. Even a "column" (better to say, a disk) 1.6 mm in length and 4.1 mm in diameter was effective in separating ribonuclease A, cytochrome C, and ovalbumin. Of course, the sample capacity of such a short column was very low [73].

4.1.8 Column Stability

Mobile phases with pH values above 7 are dangerous for silica-based columns. Even acidic aqueous solvents are reported to dissolve unprotected silica. Column life can be prolonged by the use of guard columns for presaturation of the eluent with silica. Another approach to improve stability is covering the core material with a tight hydrophobic stratum that supports the chromatographically active ligands and protects the silica against the aqueous eluent [74].

HPLC packings of higher chemical stability are available through the use of alumina. Depending on the pH of the eluent, the mechanism of the separation can be either size exclusion or ion exchange retention because the charge of the alumina surface varies with pH. The separation should be performed at a pH well above the zero point of surface charge. Here, proteins with a pH higher than the actual pH of the mobile phase will be retained by cationic exchange, whereas others with pI < pH are separated by SEC [75].

Silica-based ether-bonded phases for high-performance hydrophobic interaction chromatography proved stable at pH 6. A column used for a five-month period yielded retention data that within 5% equaled that from a fresh column immediately after the "condition period".This term refers to the first protein injections on a freshly made column with ether-bonded phases, where the protein peak height increases to a constant value with repeated injections [56]. On a new reversed-phase C 3 column, hydrophobic peptides were much more tenaciously retained than on a repeatedly used one of the same brand. This was observed with columns of 75 × 4.6 mm; d_O = 30–55 nm; d_P = 5 µm [69].

4.1.9 Predictability of Protein Retention

The chromatographic behavior of oligopeptides not exceeding 20 residues can be estimated from contributions of their constituents. In contrast to this, the experiences gathered so far suggest that each large protein requires its own specific conditions for elution. As yet, the rules for forecasting these conditions are not known. This is at least partly due to the secondary and tertiary structure of the proteins that allows only a few of their constituents to be exhibited in the outer structure of the molecule. In liquid chromatography of native proteins, this surface determines the interactions with the stationary phase. If unfolding occurs, the chromatographic behavior will change drastically. More groups can interact with the stationary phase leading to prolonged retention. If both native and unfolded conformation can exist under the conditions of liquid chromatography, double peaks occur, *vide supra*.

4.1.10 Selectivity and Resolution Power

Well-conditioned RPC can separate proteins that differ in rather small structural elements. The vasopressin-related neurophysin II, a polypeptide of 95 residues, could be separated into two isoforms that differ from each other only by one residue, either Ile or Val, at position 89. This was possible by RPC on C 8 or C 18 columns (250 × 4.6 or 150 × 4.6 mm, respectively) with rather shallow gradients (6 % B in 30 min or 20 % B in 60 min). The gradient components were (triethyl ammonium phosphate buffer, pH 3)/acetonitrile [76].

Murine epidermal growth factor is a polypeptide containing 53 residues. Its molar mass is 6040 g/mol. Chromatography on diethylaminoethyl cellulose produced a single peak. The isocratic RPC of this seemingly homogeneous material on a C 18 column (300 × 7.8 mm) yielded two clearly separated peaks (Fig. 19) [77]. The eluent was water — acetonitrile (74:26), 0.04 *M* in triethylamine acetate, pH = 5.6. The structure of the β component is identical with that of the α component but does not possess the Asn residue at chain position 1.

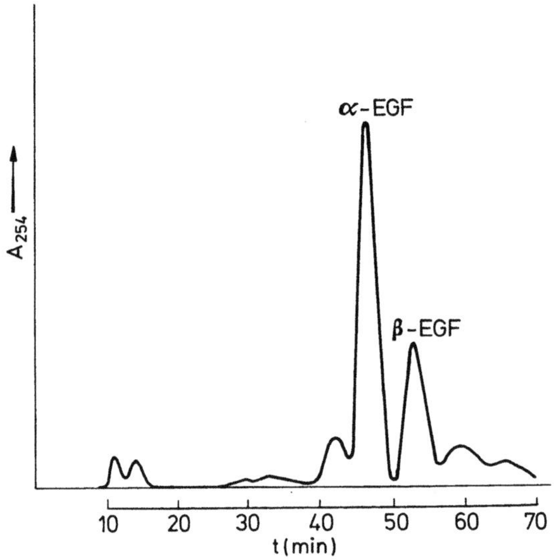

Fig. 19. Efficiency of RPC in protein separation. Isocratic RPC separation of murine epidermal growth factor (after elution as a single peak from a diethylaminoethyl cellulose column). Column: RP 18 (300 × 7.8 mm; d_p = 10 μm); 40 °C; eluent: 0.04 *M* triethylamine acetate in water — acetonitrile (74:26), pH 5.6; flow rate: 0.8 ml/min; UV detection at 254 nm, 0.05 AUFS; injection: 100 μg in 100 μl. The α peak was due to the polypeptide of 53 amino acid residues (M = 6040 g/mol), the β component did not contain the asparagine residue in position 1. (From Ref. [77] with permission)

Isocratic RPC was also successful in separating the position isomers of monoiodinated insulin. The 125-iodination by the lactoperoxidase method was carried out in phosphate buffer containing 6 *M* urea and led to the iodination of tyrosyl units, i.e., to substitution in the positions A14, A19, B16, or B26. The retention of the monoiodinated insulin chains increased in the order A19 ≪ B26 < B16 < A14 on a C 18 column (150 × 4.0 mm; d_p = 3 μm). The eluent contained 1 % triethylammonium trifluoroacetate in water (pH 3) and acetonitrile [78].

Carefully balanced HIC conditions enabled even cells to be separated under preservation of their viability. Differentiation was possible on agarose gels supporting ligands with 2–6 C atoms. Desorption with high yield in viable cells was obtained through displacement by bovine serum albumin (BSA) in phosphate-buffered salt solution at pH 7.4. The mechanism is understood in that BSA prevents re-adsorption of the cells by competing for the deserted alkyl chains [79].

Selectivity and resolution in RPC are influenced by pore size and ligand density. Hydrophilic solutes with low retention require packings with a high carbon load, whereas strongly hydrophobic samples would be better separated on a column with short ligands of low density, i.e., with a low carbon load. Obviously, the interactions between solutes and stationary phase must be balanced to the degree that they suffice for retention but do not much exceed this level. Engelhardt et al. observed better separation of protein test mixtures on a Si-500 C 18 column ($d_O = 50$ nm) than on a small-pore C 18 material. Interestingly, the resolution of a Si-100 C 18 column with 3.9 μmol/m^2 was worse than that of a related column based on Si-60 [15].

4.1.11 Separation and Reconstitution of Ribosomes, a Demonstration of the Capability of RPC in Bio-Engineering

The ribosomes of *Escherichia coli* ($M = 2.3 \times 10^6$ dalton) consist of two dissimilar subunits. The larger one is the 50 S unit containing 32 unique proteins (L1–L34) and one molecule each of 5 S and 23 S rRNA. The smaller subunit has a sedimentation coefficient of 30 S and contains one molecule 16 S rRNA and 21 unique proteins (S1–S21). The proteins of both subunits could be separated by RPC on a C 18 column (250 × 4.6 mm; $d_O = 30$ nm; $d_P = 10$ μm) by a gradient water/2-propanol with 0.1% trifluoroacetic acid in both solvents (Fig. 20). With 2-propanol, 15 proteins of the

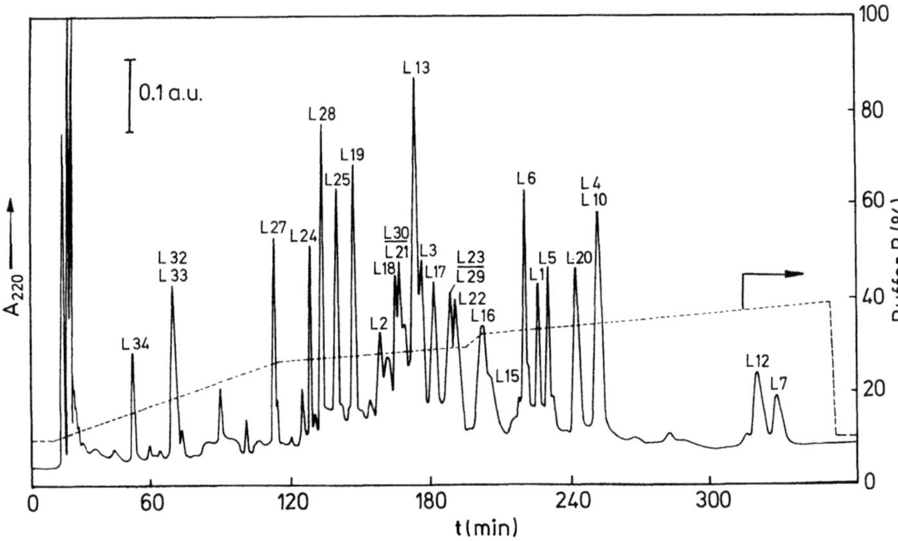

Fig. 20. Gradient elution of 50 S ribosomal proteins from *E. coli* on a RP 18 column (250 × 4.6 mm; $d_O = 30$ nm; $d_P = 10$ μm) through water/2-propanol with 0.1% TFA. Gradient program: 10% B (0 min), 27% (100 min), 30% (180 min), 33% (185 min), 38% (355 min); flow rate: 0.5 ml/min; injection: 2 mg TP 50 in 200 μl 2% acetic acid. (From Ref. [80] with permission)

30 S subunit were obtained of sequencer purity [80]. (Acetonitrile as modifier yields this purity level for only 4 proteins.) The aim of this work was isolating and sequencing the various ribosomal proteins. A list of all 52 proteins shows molar mass values in the range from 5381 (46 residues) to 61,159 g/mol (557 amino acid residues) [80]. The proteins L26 and S20 are identical.

By another team, the separations were performed on a RP18 column (250 × 4.1 mm; d_O = 30 nm; d_P = 6.5 μm) through a gradient water/acetonitrile, both solvents with 0.1% TFA.

The goal of this work was to reconstruct the subunits from their separated components [81]. Reconstitution after storing in 0.1% TFA at −80 °C for up to six months produced protein combinations that contained stoichiometric amounts of the constituents and yielded chromatographic traces similar to those of the starting material (Figs. 21 a, b). After reconstitution with the respective rRNAs, the activity of the 30 S

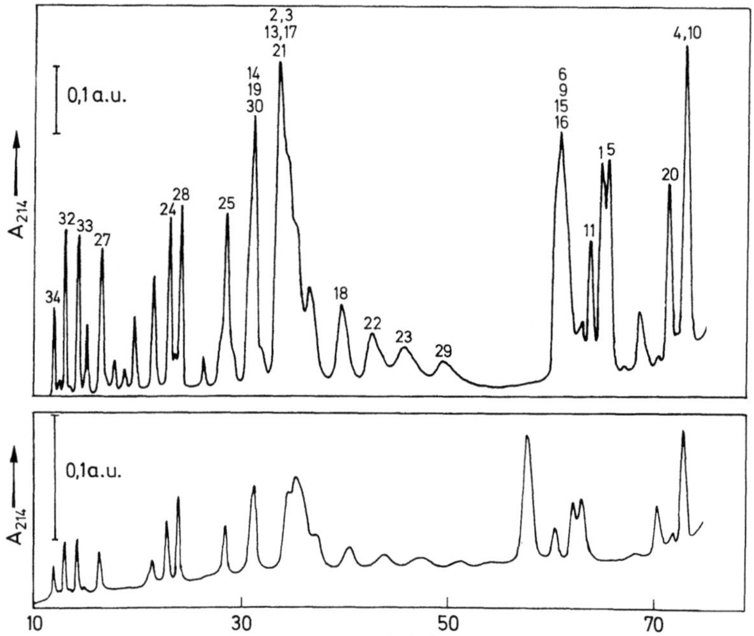

Fig. 21 a and b. Gradient elution of 50 S ribosomal proteins from *E. coli* on a RP 18 column (250 × 4.1 mm; d_O = 30 nm; d_P = 6.5 μm). Gradient: water/acetonitrile, with constant admixing of 10% (v/v) water + 10 g/l TFA; gradient program: 17% AcN (0 min), non-linear to 34.9% AcN (20 min), 36% (30–40 min), 45% (50 min), 50% (55 min), 75% (60 min), 75% (65 min); flow rate: 0.7 ml/min; UV detection at 214 nm. *a* = Injection of 700 pmoles TP 50 in 60 μl 0.1% TFA, 0.6 AUFS; *b* = injection of 180 pmoles of reconstituted 50 S subunit, dissolved in 0.1% TFA, 0.15 AUFS. (From Ref. [81] with permission)

subunit was monitored by the PhetRNA^{Phe} binding and that of the 50 S subunit with a peptidyltransferase assay. About 30% activity was obtained for the 30 S subunit at a protein to rRNA ratio of 1.8:1, and a maximum of 84% at a ratio of 3.6:1 [81].

The ultimate goal of these studies was performing single replacement experiments, e.g., by reconstituting subunits containing puromycin-labeled S14 or L23 protein in place of unmodified S14 and L23.

4.2 Separation of Nucleic Acids and their Fragments

HPLC separation of DNA fragments is an important contribution to DNA research and gene technology. Progress in restriction enzymology, gene cloning, and DNA sequencing demanded a separation technique capable of isolating milligram amounts of pure DNA fragments for biochemical or physical investigations.

On a small scale, DNA restriction fragments can be separated and characterized by polyacrylamide gel electrophoresis, but this analytical technique fails in certain cases as, e.g., for fragments of similar chain length even if differing in base composition or end-group structure. HPLC is capable of separating such components and has further the unique capacity of isolating the complementary strands of DNA restriction fragments. It is also an efficient tool for preparing pure and well-defined oligonucleotides from synthesis or degradation mixtures. Further applications of HPLC in DNA research will be the separation of DNAs slightly differing in non-paired nucleotides, such as growing fork regions in DNA replication, the separation of covalently closed DNA rings with different numbers of supercoiled turns, and the separation of protein/nucleic acid complexes from bare DNA [82].

Potential packing materials for nucleic acid chromatography have hydrophobic and weakly cationic sites. Figure 22 shows the logarithmic retention factors of oligoadenylyl phosphates on an ion-exchange column (100 × 4.6 mm) with ethylene di-

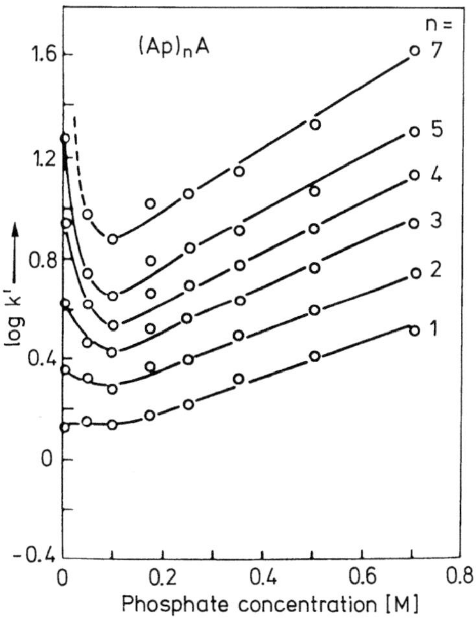

Fig. 22. Mixed mode retention. Logarithmic retention factor for oligoriboadenylic acids vs. phosphate concentration in the mobile phase, pH 6.3. The figures at the curves indicate the number of adenylyl phosphate residues in the samples. (From Ref. [74] with permission)

amino tetra(2-propanol) bound to the sublayer of C 3 hydrocarbon chains. The supporting material was spherical silica gel, $d_O = 33$ nm, $d_P = 5$ µm, the carbon load 2.56% (w/w). The abscissa scale indicates the phosphate buffer concentration in the mobile phase (pH = 6.3). At a low electrolyte concentration, the retention decreased with increasing salt concentration; this was indicative of retention by an anionic exchange mechanism. The increase of log k′ at higher ionic strength was due to hydrophobic interactions [74]. A similar behavior was observed with supports containing modified amino groups: decaying k′ values of oligo-nucleotides with increase in pH or acetonitrile content of the eluent indicated a mixed mode retention, i.e., contributions from electrostatic and hydrophobic interactions [83].

A column (250 × 4.6 mm) packed with a reversed-phase C 18 support, coated with trioctylmethyl ammoniumchloride, separated oligo-uridylic acids and discriminated between $(Up)_{89}$ and $(Up)_{90}$, (Fig. 23)[83]. In such high a region of molar mass, the separation of components differing in only one monomeric unit is remarkable in polymer chromatography in general.

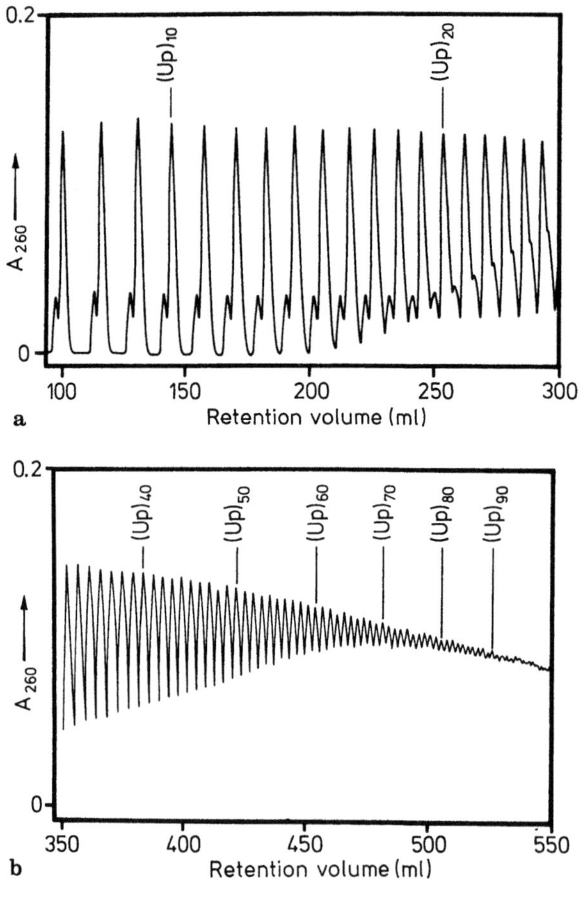

Fig. 23a and b. Gradient elution of oligouridylic acids on a RP 18 column (250 × 4.6 mm; d_P = 5 μm). Gradient component A = 0.5 *M* ammonium acetate, pH 4.5, component B = 5.0 *M* ammonium acetate, pH 6.0; gradient program: from 0 to 40% B in 18 h; flow rate: 0.5 ml/min. (From Ref. [83] with permission)

4.3 Separation of Synthetic Polymers

4.3.1 Oligomers

Low-molecular polystyrene standards (600 and 900 g/mol) were isocratically separated into individual oligomers on a C 18 column (250 × 3.2 mm; d_O = 6 nm; d_P = 7 μm).

The eluent was pure acetonitrile. The number of isolated peaks was 10 or 11, respectively, with some indication of a fine structure [84].

Repetition of this work with a similar C 18 column (250 × 4.0 mm; d_O = 6 nm; d_P = 5 μm) clearly showed the presence of stereoisomers in oligomers with 4–8 repeat units, separated from a PS 600 standard [14], but the resolution was not as good as in comparable work on plain silica [46], *vide supra*.

Gradient elution of polystyrene oligomers was performed with methanol/tetrahydrofuran systems. In the first paper, a C 18 column (150 × 4.6 mm; d_P = 6 μm) was used for the separation of a PS 800 sample. The gradient rate was 2.1% THF/ml at 1.5 ml/min flow rate (S. D. Abbott in [85]). In another investigation, the sample was PS 750, the column C 18 (300 × 3.9 mm), the gradient rate 1.67% THF/ml, and the flow rate 1 ml/min. Here, the peaks were characterized by mass spectrometry [86]. Interestingly, neither of these separations indicated any substructure.

Phenyl-oligo(ethylene glycol) homologs were separated into individuals on a C 8 column (250 × 4.6 mm; d_O = 6 nm; d_P = 7 μm) with a water/acetonitrile gradient (from 20 to approximately 33% B in 30 min) at 2.0 ml/min flow rate [87] (Fig. 24).

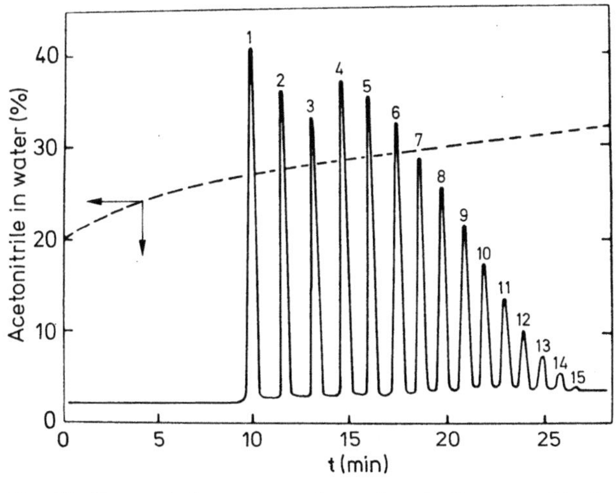

Fig. 24. Reversed-phase chromatography of oligomers. Gradient elution of phenyl-oligo(ethylene glycol) homologs on a RP 8 column (250 × 4.6 mm; d_O = 6 nm; d_P = 7 μm). Gradient: water/acetonitrile; program as indicated; 25 °C; flow rate: 2 ml/min; UV detection at 254 nm. (From Ref. [87] with permission)

A commercial poly(tetramethylene ether)glycol with a molar mass of 1000 g/mol was UV-labeled with phenylurethane endgroups and separated on a C 8 column (250 mm in length) with the help of a water/methanol gradient (85–92.5% B in 15 min and from 92.5 to 100% B in further 50 min). The flow rate was 2 ml/min [88].

Epoxy resin prepregs were characterized by RPC on C 18 columns with a water/tetrahydrofuran gradient, either linear from 40 to 100% B within 30 min or concave from 40 to 80% B in 50 min [89]. Reversed-phase chromatography revealed more details than the SEC of the sample. Samples that looked similar in SEC showed marked differences in RP-HPLC (Fig. 25).

The acetone extract of an SBR compound was analyzed on a C 18 column through

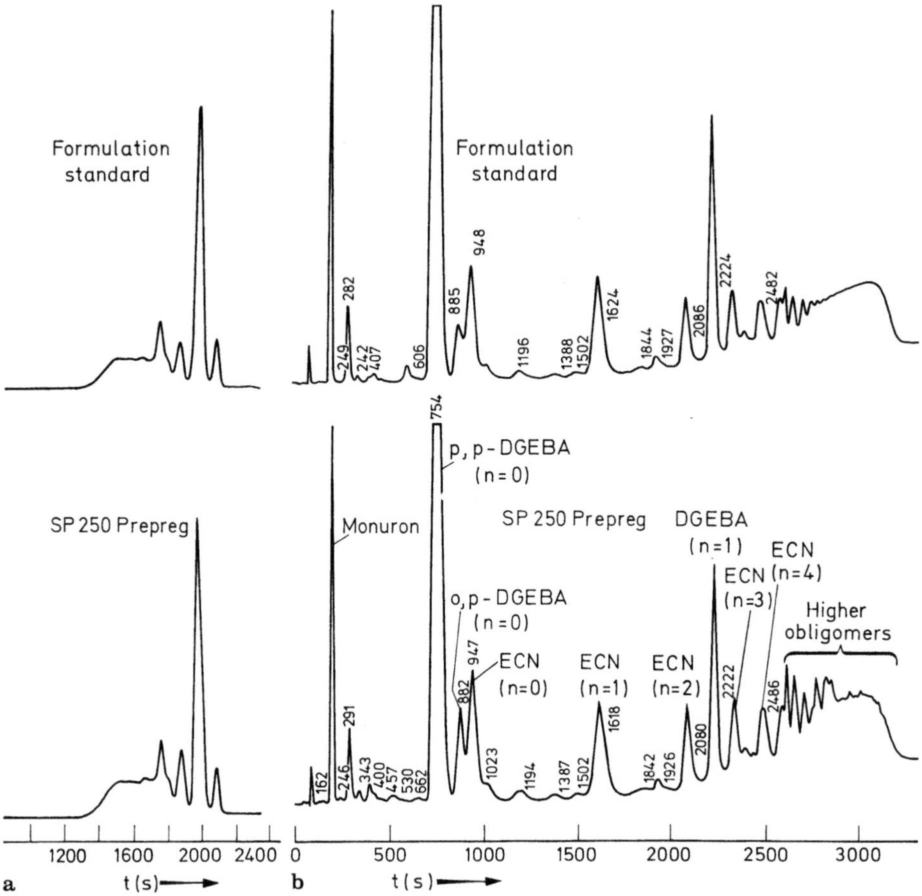

Fig. 25 a and b. Analysis of epoxy resin prepregs. **a** = SEC with UV detection at 280 nm, comparison of sample SP 250 with formulation standard. **b** = Gradient elution of these samples on a RP 18 column with water/THF; gradient program: 40 to 80% B in 50 min, non-linear. (From Ref. [89] with permission)

a water/(tetrahydrofuran-acetonitrile) gradient from 35 to 80% THF and from 10 to 15% AcN within 25 min. With fluorescence detection (230 nm excitation, 400 nm emission), numerous peaks could be seen in the chromatogram [90].

The investigations mentioned so far aimed at objectives of rather low molar mass. Reversed-phase chromatography of polystyrenes with 17,500 and 50,000 g/mol was performed on C 18 columns with water/tetrahydrofuran gradients or mixtures [3]. The latter sample was isocratically eluted from 30 nm-pore packing with 85% THF as a broad band, 87% THF let the polymer elute in the void volume, whereas 83% did not produce an observable band at all.

The retention factors of the polymers in 60% THF were calculated from gradient elution and discussed together with those directly measured for oligomers with 2–12 repeat units or 2000, 4000, and 9000 g/mol molar mass. On a 100nm-pore packing, the Martin equation was fulfilled for n = 19 repeat units. At higher values, the curve in the plot log k′ vs. n levels off. Even though the slope of the Martin plot decreased, the calculated k′ values of the polymers practically indicated infinite retention.

In comparison with the tremendous number of low-molecular RPC separations, the limited variety of polymer separations is astonishing. Precisely spoken, the lack of applications to synthetic polymers is strange. The great potential of RP-HPLC for the separation of biopolymers (*vide supra*) has been stressed with the intention of making this point very clear. The number of HPLC investigations of synthetic polymers by normal-phase chromatography is not to such a degree disappointing. Here, a polar stationary phase attracts the solute from a solution in less polar eluents. In normal-phase chromatography, the problem eventually might be how to remove the retained polymer from the stationary phase. Several successful separations demonstrate that this obstacle can be overcome.

In solvophobic retention, the surface of the column packing is more or less passive. Silanophilic interactions are known to occur in water-lean eluents, but the bulkiness of macromolecules render them unable to reach the remote silanols. Even rather small models, such as heptapeptides showed no silanophilic retention when both the C and N terminus had been blocked [91]. If the COOH and/or the NH_2 terminals were free the model peptides showed the usual behaviour of polypeptides and proteins: The k′ values declined with increasing volume fraction φ of organic modifier, passed a minimum at about $\varphi = 0.6$–0.8 and rose again at high φ values. This is due to hydrophobic interaction at low φ and silanophilic retention at high φ values. Synthetic polymers prepared by common polymerization techniques should not have endgroups of high polarity. Thus, for polymers, solvophobic retention remains the predominant mechanism of reversed-phase chromatography. It requires the mobile phase to be more polar than the sample and the stationary phase. Proteins and polypeptides are amphophilic, hence, aqueous eluents can induce hydrophobic interactions and direct the solutes towards the stationary phase.

The solubility of polymers is, for thermodynamic reasons, more restricted than the solubility of low-molecular compounds and, consequently, the choice of solvents is limited. Potential solvents for most synthetic polymers are of moderate polarity. Alcohols and liquids of similar polarity are precipitants for many synthetic polymers. The search for a mobile phase that enables RPC through solvophobic interactions between the polymer and the nonpolar stationary phase requires attempts to make the mobile phase an unfavorable environment for the solute. This easily conflicts with the narrow limits of solubility of the polymer under investigation. Solubility effects are known to occur even in low-molecular RPC [92–94], but in polymer RPC they even may govern retention.

5 Separation by High-Performance Precipitation Chromatography

The precipitation of a polymer can be regarded as a solvophobic phenomenon: the interactions between polymer and solvent molecules are diminished by lowering the temperature or adding nonsolvent until they are weaker than the intra- and intermolecular interactions between polymer segments. If the thermodynamic quality of the solvent has decreased beyond a certain limit, the macromolecules associate and eventually coalesce into a polymer-rich gel phase. The gel phase is in equilibrium with the surrounding polymer-lean sol phase and contains a higher concentration of the favorable solvent than the latter.

5.1 Baker/Williams Fractionation

Fractionation by means of precipitation chromatography was suggested by Baker and Williams [95]. The method has been reviewed recently [96,97]. Numerous applications to copolymer investigation have been reported [98–107].

Poly(styrene-*co*-acrylonitrile) samples (SAN) have been fractionated through methanol-acetone or methanol-dichloromethane mixtures of stepwise-altered composition [108] through methanol/dichloromethane, *n*-hexane/dichloromethane, or methanol/acetone gradients [109], cyclohexane/methyl ethyl ketone gradients [110], or (toluene — 1-propanol 50:50)/dimethyl formamide gradients [111]. Mixtures of ethylene cyanohydrin — ethylene carbonate [112] or cyclohexane — methyl ethyl ketone [113] have also been used for the separation of SAN copolymers. Both the latter systems as well as the *n*-hexane/dichloromethane gradient [109] had the efficacy of fractionating SAN copolymers by composition.

5.2 High-Pressure Precipitation Chromatography

Attempts to utilize the efficiency of modern HPLC techniques for separating SANs by composition were successful with *n*-hexane/tetrahydrofuran gradients [114]. The surprising observation that the separation could be obtained on a silica column (250 × 4.6 mm; d_O = 10 nm; d_P = 10 µm) as well as on a RP8 column (250 × 4.6 mm; d_O = 10 nm; d_P = 10 µm) was a challenge for further investigation. The eluent com-

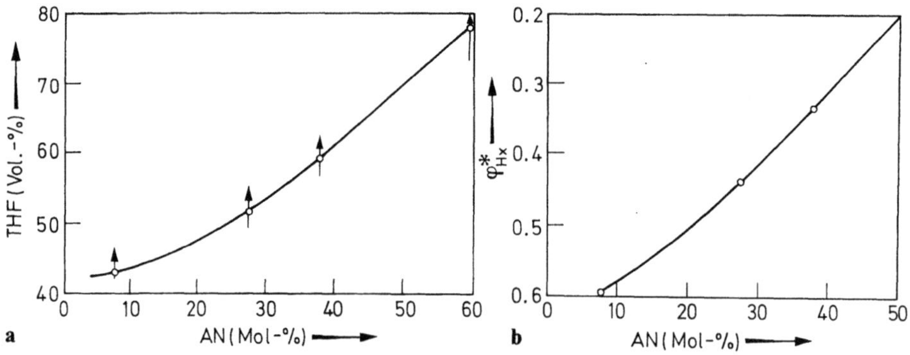

Fig. 26a. Elution characteristics of poly(styrene-*co*-acrylonitrile) samples. % THF vs copolymer composition (mol % acrylonitrile) from gradient elution on a silica column (250 × 4.6 mm; d_O = 10 nm; d_P = 10 µm) at 40 °C. Gradient: *n*-hexane/THF; program: 20% B (0 min), 35.6% (2 min), 90% (16 min). Injection: 20 µg in 50 µl THF. The arrows indicate the band width. **b** = Solubility curve of SAN copolymers in *n*-hexane — THF mixtures from turbidimetric titrations. The points were measured by investigation of each sample dissolved in THF (c_0 = 40 mg/l) through stepwise addition of *n*-hexane at 20 °C. Both diagrams conincide in scale. (By courtesy of Akademie Verlag, Berlin [114])

ponent *n*-hexane is a nonsolvent for SAN copolymers. By comparison with results from turbidimetric titrations, it became clear that the copolymers of different composition were eluted just at the verge of precipitation (Fig. 26). In another approach to separate these samples, a CN column (50×4.0 mm; $d_O = 6$ nm; $d_P = 10$ μm) was used with a gradient starting with a 70:30 (v/v) mixture of heptane (nonsolvent) and dichloroethane. The solvent B was acetonitrile — dichloroethane (40:60). The gradient was linear from 3 to 100% B within 13 min [115].

An independent approach to precipitation-controlled HPLC (HPPLC) was the nonaqueous reversed-phase liquid chromatographic fractionation of polystyrene: the mixture of 7 polystyrene standards with molar masses ranging from 2350 to 10^7 g/mol was separated by molar mass on a C 18 column through a methanol/dichloromethane gradient at 2.5%/ml gradient rate and 1 ml/min flow rate. The starting concentration was 53% B [116]. The polarity order (A: polar, B: less polar solvent) was within the framework of RP chromatography, but the component A was a nonsolvent. Additional experiments were performed on reversed-phase TLC plates. From suspected eluent demixing on the TLC it was concluded "that one might have in principle a separation akin to the fractional precipitation of polymers". The mechanism was confirmed by theoretical considerations [117] and diagonal spot TLC [118].

As always in gradient elution, the gradient program influences the shape of a HPPLC chromatogram. A linear increase in % B with time at a constant flow rate facilitates quantitative evaluation, but a more sophisticated program may be advantageous for obtaining linear relations between retention and the sought quantity of the sample polymer, i.e., a linear calibration.

If retention is only due to the solubility mechanism, the initial conditions of analyses must be such that the polymer precipitates from the injected solution. In general, the sample is injected at ambient temperature. If the possibly differing column temperature does not cause precipitation, the latter must be obtained by (partly) exchanging sample solvent for nonsolvent from the "running" eluent. This should happen in the uppermost part of the column. Such a process is straightforward if narrow-pore packings are used and the initial eluent is a nonsolvent. Provided that the polymer is too bulky for penetration into the pores, it is restricted to the interstitial channels, whereas the solvent from the sample can exchange freely with the nonsolvent in the pores.

The interstitial volume V_I of the column is the accessible volume for an excluded polymer. If dissolved, the polymer is transported at a linear velocity:

$$u_P = L/(V_I/v) \tag{10}$$

Here, L is the length of the column and v the flow rate.

In contrast to the polymer, the eluent has access to the pore volume of the packing. The eluent volume in the column is V_m. This gives a linear velocity for the eluent:

$$u_E = L/(V_m/v) \tag{11}$$

As $V_m > V_I$, the velocity of the eluent is smaller than the velocity of the dissolved polymer. The polymer bypasses the pores and thus always overtakes the eluent having a sufficient solvent power to keep the polymer in solution. It rushes into the poorer solvent running in front, and consequently precipitates. This gives the opportunity for multi-stage separation [119].

This model has been recently quoted[1] in some discussions [120–122]. To avoid misunderstandings, it might be stressed that a precipitation mechanism was proposed [114] on the base of

- the parallelism between elution and solubility (cf Fig. 26)
- the successful separation by the combination of reversed-phase columns with a normal-phase gradient
- the tolerance of the peak position to changes in column parameters (cf. Table 1, *vide infra*)
- the molar-mass dependence of the retention identical to that of solubility.

Table 1. Solvent composition at sample elution (volume percent of eluent component B)

Column	SAN sample (wt % acrylonitrile)				
	16.1	23.0	29.1	36.4	42.5
LiChrosorb SI 4000 RP 18, d_P 10 μm, 150 × 4.6 mm	58.1	66.8	74.1	80.7	91.2
Polygosil 60-5 C_{18}, d_P 5 μm, 150 × 4.6 mm	57.0	66.5	74.0	80.7	90.4
μBondagel E-1000, d_P 10 μm, 300 × 4 mm	57.1	66.3	73.8	80.3	88.5
Nucleosil 5 CN, d_P 5 μm, spherical, 150 × 4.6 mm	57.7	66.7	74.1	80.6	89.1 +91.3
	57.7	—	73.8	—	nr
	—	66.5		80.4	—
LiChrosorb Diol, d_P 10 μm, 150 × 4.6 mm	excl	—	73.3 +74.8	—	89.6
	excl	67.6	73.0 +74.8	79.6 +81.5	90.0
Polygosil 60-5, d_P 5 μm, 150 × 4.6 mm	57.2	66.3	73.8	80.3	91.1
Polygosil 60-5, d_P 5 μm, 55 × 4.6 mm	57.1	66.1	73.5	80.4	91.1

nr: no resolution (sample eluted during gradient return);
excl: exclusion (sample eluted before the solvent THF);
+: two figures for one sample indicate peak splitting

The model is not needed and was not used as a proof for a HPPLC mechanism. It was used in order to point to the opportunity of multi-stage retention which is unfortunately not that straightforward with other types of gradient elution.

If the packing does not effect any attraction towards the sample and the starting eluent is not sufficiently poor in a thermodynamic sense, the polymer will elute un-

1 The author is very much indebted to Dr. L. Snyder for making available the manuscripts of Ref. 121 & 122.

retained in the interstitial volume of the column. For example, SAN copolymers could be separated by composition through a *n*-hexane/tetrahydrofuran 'gradient in the range of about 40–80 % B. Retention was positively obtained on a C 8 column (250 × 4.6 mm; d_O = 10 nm; d_P = 10 μm) at 40 °C if the injection was performed into 20 % THF. Preliminary experiments with 40 % THF in the starting eluent produced, in some cases, unretained elution of a polymer with 16.1 wt-% acrylonitrile [114].

The gap in eluent composition between the precipitating initial concentration of B and the separating range can be passed rapidly. Since some UV detectors answer a swift change in refractive index with a pseudopeak, the refractivity of the gradient components should not differ too widely.

Ideally, the sample solvent should be identical with the eluent running at injection. For the retention requirements just discussed this cannot be fulfilled in HPPLC. Any injection solvent differing from the eluent causes overload with respect to the former. Methyl ethyl ketone as a sample solvent produced a large UV signal after injection of 20 μl into an *iso*-octane/THF gradient. At 1 ml/min flow rate, it took about 4 min until the MEK was eluted from a C 18 column (150 × 4.6 mm; d_O = 6 nm). Thus, transparency of the sample solvent at the detection wavelength is at least required, but the sample solvent should better be similar to the starting eluent in chromatographic quality as well. Even when invisible, the solvent plug might disturb column conditioning. THF (with butylated hydroxy toluene stabilizer added) proved suitable as a sample solvent if *iso*-octane/THF gradients were used with low initial concentration of THF. In hundreds of gradient runs with polymer retention times of at least 4 min, disturbancies were never observed. The sample solvent produced a sharp signal approximately at the mobile phase volume of the column. Its appearence announced the performance of an injection even before the sample was eluted.

Two advantages of HPPLC should be mentioned here: its general applicability and its only moderate dependence on column activity. The first aspect refers to the fact that solubility and precipitability are common characteristics of a wide variety of polymers. In principle, HPPLC is Baker-Williams fractionation by means of modern HPLC techniques.

If the polymer absorbs at a wavelength where both solvent and precipitant are transparent, HPPLC can be performed with the help of a UV detector. Styrene-containing copolymers can be analyzed with alkane hydrocarbon/THF eluents at around 254 nm. In order to gain high sensitivity in gradient elution, we removed the stabilizer from the THF by distillation under N_2 [123]. Caution is required with respect to peroxides.

At 259 nm, even freshly distilled THF is less transparent than *iso*-octane. This will cause baseline rise of a sensitive UV detector which might impair peak detection. Levelling can be obtained by adding a UV-absorbing but chromatographically inert concomitant, e. g., 10 ppm toluene, to the alkane hydrocarbon.

Figure 27 shows a typical gradient for separation of SANs by composition together with a corresponding detector signal. In order to adjust the refractive index of mobile-phase component B close to that of component A (*iso*-octane, n_D = 1.392), a mixture of THF (n_D = 1.405) with 10 % methanol (n_D = 1.329) was used as solvent B. The second advantage of HPPLC is the restricted influence of column packing material. Figure 28 shows two almost equal chromatograms obtained either with a C 18 (upper) or silica-packed column (lower curve) [124]. In Table 1 the THF contents are listed

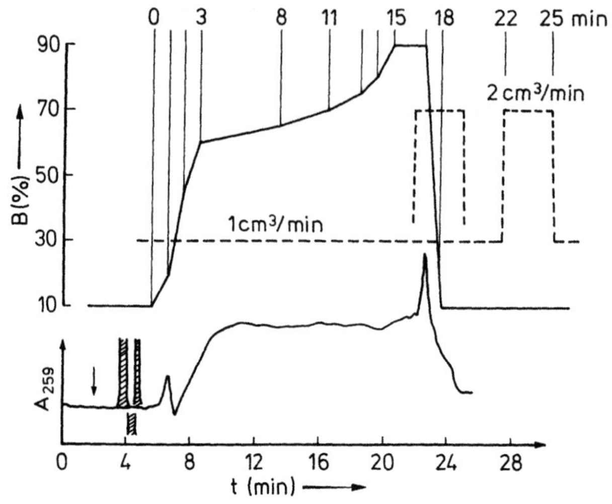

Fig. 27. Gradient and flow-rate pattern (upper diagram) and the corresponding UV report (blank) of a gradient cycle used for the HPPLC separation shown in Fig. 31., e.g., Gradient: *iso*-octane/(THF + 10% methanol). Detector: Schoeffel type SF 770 UV detector, at 259 nm wavelength. The shaded inflections indicate the position where a solvent peak would occur on sample injection. The shift in time scale is due to the lag time of the system (5.6 min)

which where observed at peak elution. Through these figures results can be compared that were obtained on a variety of columns even with differing gradient programs. The roughly constant values from strongly dissimilar columns are surprising, but it must be stressed that in some experiments distorted or even split bands had appeared, especially with the diol column.

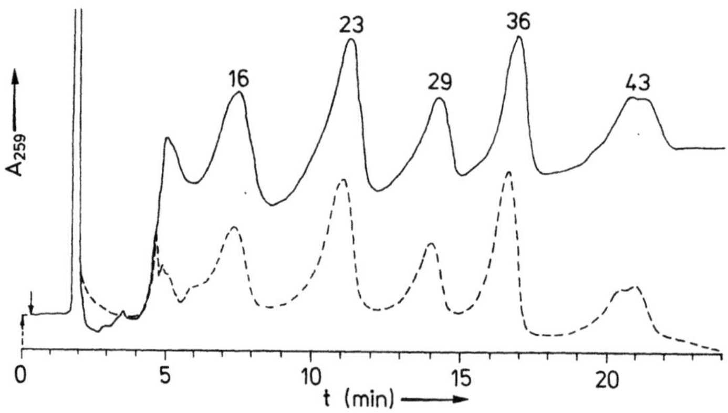

Fig. 28. Gradient elution of the mixture from five SAN copolymers on either a RP 18 column (150 × 4.6 mm; $d_O = 6$ nm; $d_P = 5$ μm, upper chromatogram) or a silica column (150 × 4.6 mm; $d_O = 6$nm; $d_P = 5$ μm, lower chromatogram). Gradient: *iso*-octane/THF; program: 10% B (0 min), 50% (2 min), 100% (22 min); temperature: 50 °C; UV detection at 259 nm. (Note the time shift due to a small difference in lag time). The figures at the peak indicate the acrylonitrile content (w/w) of the specimens [124]

The practical consequences of the stability against changes in column activity are

- long life-time of columns (a C 8 column worked well over a period of about two years);
- less stringent requirements for column conditioning than in adsorption LC;
- the chance of successful separation even in the presence of concomitants.

The influence of pore size on the elution of SAN samples was small. Short columns yielded better separation than longer ones. This is similar to a recent observation in protein RPC [73], *vide supra*. Figure 29 shows that a 55 mm column was superior to a 150 mm one in separating a mixture of five poly(styrene-*co*-acrylonitrile) samples. (Note that the leading pseudopeak was also diminished.)

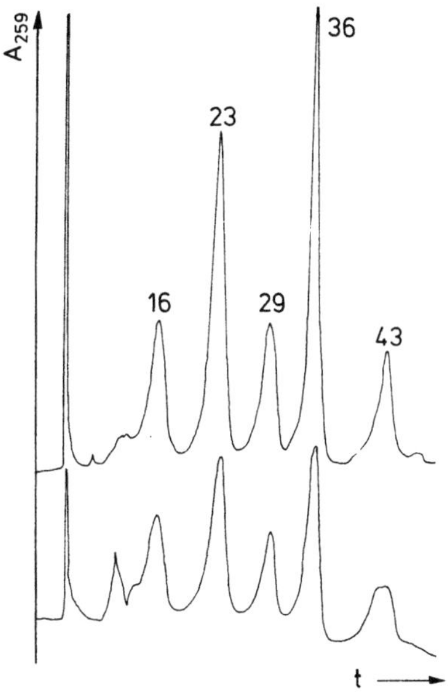

Fig. 29. Influence of column length on gradient elution of SAN copolymers through *iso*-octane/THF. Column packing material: Polygosil 60-5 silica in both cases, $d_O = 6$ nm; $d_P = 5$ µm. Upper chromatogram: column 55 × 4.6 mm, lower one: column 150 × 4.6 mm. (The lower chromatogram is from the same run as the dashed line in Fig. 28.) Injection: 60 µg in 30 µl THF [124]

Higher column temperature should improve separation due to accelerated diffusion. The retention decreased with increasing temperature which is in accordance with increasing solubility. (In reversed-phase chromatography retention also decreases with rising temperature.) HPPLC must be carried out with temperature control. The chromatograms shown in Figs. 28 and 29 were obtained at 50 °C. The gradient in HPPLC must be one of increasing thermodynamic quality. For instance, the separation shown in the upper curve of Fig. 28 was accomplished with *iso*-octane/tetrahydrofuran. This is a gradient increasing in polarity as it would be necessary for adsorption chromatography on a polar packing — but the column was packed with C 18! This implies that the separations observed were not due to a classical mechanism.

It is well known that the solubility of polymers depends on their molar mass. This effect can be utilized for the evaluation of molar-mass distributions. With copolymers, solubility is determined by composition and molar mass. Topčiev et al. derived the equation:

$$\varphi''_{P,x}/\varphi'_{P,x} = \exp P(\sigma + Kx) \tag{12}$$

Here, $\varphi''_{P,x}$ and $\varphi'_{P,x}$ are the volume fractions of the macromolecules, with the degree of polymerization P and the composition x, in the gel phase and in the sol phase, respectively; x is the volume fraction of A units in an A—B — copolymer, and K is a quantity determined by the differences in interaction parameters of solvent and copolymer constituents [125].

Equation (12) implies that the values of σ and K determine whether separation by molar mass or composition would prevail. At best, a large K will result in a good approach to separation by composition, but there is not any combination of K and σ that can totally suppress a simultaneous separation by molar mass.

6 Chromatographic Cross-Fractionation

The term cross-fractionation (CF) refers to analyses of distributions in differing directions by means of separation processes. Cross-fractionation is a significant tool for the evaluation of the complex distribution which copolymers normally have with respect to molar mass (MMD) and chemical composition (CCD). The idea of CF implies separation by one parameter and subsequent analysis of the fractions obtained for the distribution of the other parameter through another separating process.

With the classical fractionation technique, CF requires subdividing the sample into 5–10 primary fractions and each of them into 5–10 subfractions, with isolating and characterizing the polymer material of each final portion for molar mass, composition, and amount. This requires at least 15 g of starting material and several weeks of skillful work.

The idea of chromatographic CF is the use of timely chromatographic apparatus and technique for this purpose. Basically, it is the application towards copolymer investigation of what is known among chromatographers as multi-dimensional or coupled column chromatography.

These techniques are useful for the separation of complex mixtures. Quantitative analyses have been reported on malathion pesticide in vegetable matter, limonin in grapefruit peel, and additives in compound rubber by means of SEC and RPC [126]. SEC was performed with unstabilized THF and monitored at 215 nm. Working so near to the UV cut-off (210 nm) of THF was possible through the use of a detector with a flowing reference installed between the pump and the sample valve. Then the eluent passed the column and finally the analytical cell. Since the column back pressure never exceeded 10 MPa, this eluent line caused no problems.

For the investigation of a butadiene-styrene rubber, a set of three SEC columns in series was used: 500 × 8 mm, d_O = 150 nm; 500 × 8 mm, d_O = 25 nm; 800 × 8 mm, d_O = 4 nm, all three with d_P = 10 μm. The flow rate was 1 ml/min. The injection amounted to 200 μl of a 2% solution from which the carbon black had been removed. Although the additives of interest were separated from the polymer, they were still covered by an intense band from process oil. Hence, coupled-column chromatography with reversed-phase separation of SEC eluates became neccessary. 10 μl of the latter were injected into a C 8 column (250 × 2.2 mm; d_P = 10 μm) and analyzed at 0.5 ml/min flow rate through a water/acetonitrile gradient (rising from 20% B by 6%/ml). Here, UV detection was performed at 254 nm. The peaks of the additives could be clearly separated from the process oil band. The technique also proved useful for checking the identity of peaks occurring in SEC. The apparatus were either directly coupled (with the flow stopped in SEC during gradient RPC) or off-line operated. The latter provided the possibility of eluate preconcentration or changing the eluent.

6.1 Chromatographic Cross-Fractionation of Polyolefines

Poly(ethylene-*co*-propylene) samples were fractionated by composition through an exponential gradient (butyl cellosolve)/xylene from a column (1000 × 80 mm) packed with Celite 545. The samples were initially deposited onto the column packing by slowly cooling solutions in a mixture of the gradient components containing 30% (v/v) of the nonsolvent butyl cellosolve. The fractionation temperature was adjusted to a controlled value between 117 and 123 °C, depending on the melting temperature of the polyethylene segments in the copolymer under investigation. The fractions eluted were characterized for their MMD by means of SEC on polystyrene-gel columns in *o*-dichlorobenzene at 135 °C [127].

Samples of branched polyethylene were investigated by the combination of crystallizability fractionation and fractionation by molar mass. The first step was the precipitation of the polymer sample onto the glass-beads in a column (150 × 8 mm) used for the fractionation by crystallinity. The precipitation was performed by cooling a solution of 10 g/l polyethylene in *o*-dichlorobenzene from 140 to 40 °C within 60 min. Subsequently, the polymer was extracted from the column by the same solvent at stepwise increased temperature. In an example given, the first fraction was eluted at 40 °C and additional 17 fractions at temperatures each raised by 2–10 K. The finer steps were employed in the middle of the fractionation procedure; the last one reached even from 110 to 140 °C. About 10 min equilibration time proved to be adequate. The fractions were analyzed subsequently by SEC on a polystyrene-gel column. The whole process was automated [128].

Whereas in the example just described the sample amount was about 50 mg, a similar procedure developed by another group [129] started with 4 g polyethylene copolymer. The sample was applied as a dilute solution in xylene and precipitated by very slow cooling (1.5 K/h) onto the Chromosorb P packing of a 500 × 127 mm column. The first separation was *temperature-rising elution fractionation* at a flow-rate of 20 ml/min and a linear temperature increase by 8 K/h. The MMD of the fractions was measured by SEC at 145 °C in *o*-dichlorobenzene at 0.7 ml/min flow rate. The column set included a pair of bimodal columns 100 Å and 1000 Å plus a 4000 Å column. The apparatus was equipped with an IR detector. The experimental data is computed to show the distribution of short-chain branching and of molar mass simultaneously.

6.2 CCF of Copolymers Containing Styrene and Acrylate or Methacrylate Units

A poly(methyl methacrylate-*b*-styrene-*b*-methyl methacrylate) sample with an approximately equal content of both monomeric units was fractionated by preparative SEC into 34 fractions. The SEC eluent was dimethyl formamide and the starting amount of sample 40 mg. The polymer isolated from the SEC fractions was further fractionated by TLC on silica plates. Through chloroform-methanol mixtures as eluents, two subfractions were obtained, one remaining at the start and another one migrating to reasonable R_f values. After the TLC development, the polymer traces were extracted with acetone and investigated by pyrolysis gas chromatography. From the ratio of styrene to methyl methacrylate units in both parts as well as in the starting SEC fraction and from considerations about the migration probability of the block copolymer, the methyl methacrylate content of the latter was estimated. It decreased from 96% in SEC fraction No. 5 to 43% in fraction No. 24 [130].

Poly(styrene-*b*-methyl methacralate) samples were also cross-fractionated by column adsorption chromatography and subsequent SEC of the fractions. The initial fractionation was carried out in a column (150 × 50 mm) in a manner closely related to TLC. The sample (300 mg) was loaded on 20–30 sheets of filter paper. They were piled up in the column upon a zone of 20 mm dry silica and a glass fritt where the developer would eventually be drawn in. After compression, the impregnated sheets formed a sample bed about 10 mm in height. The chromatographic packing of activated silica was located upon the sample bed and subdivided by filter paper into 9 compartments of approximately 10 mm height each. The lower end of the column was then immersed in a mixture of benzene and ethyl acetate (27.5:72.5 v/v). The developer was soaked up by the dry packing at ambient temperature in about 5 hours. Then the column was dismanteled and the silica from each compartment extracted by THF. The fractions were investigated by SEC with THF as an eluent. Figure 30 shows the result of the chromatographic cross-fractionation which disclosed the chemical composition

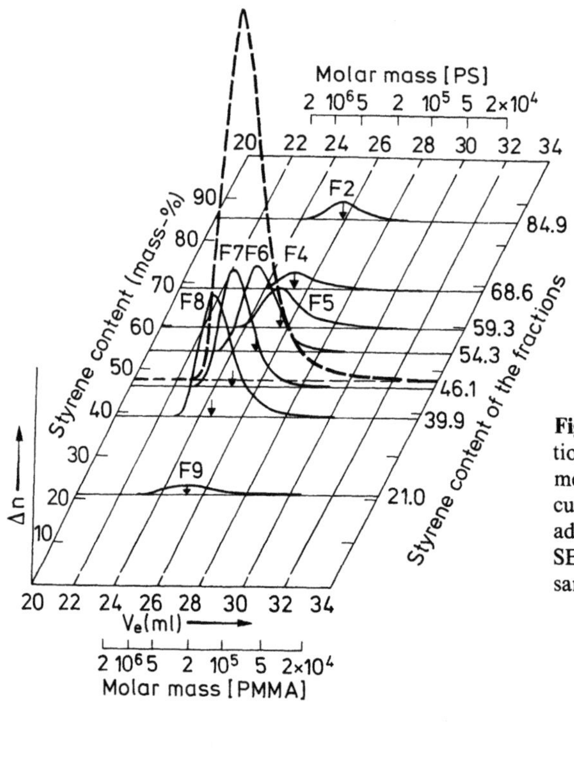

Fig. 30. Chromatographic cross-fractionation of 300 mg poly(styrene-*b*-methyl methacrylate) with 47 wt % styrene. SEC curves of the fractions obtained by column adsorption chromatography. The large SEC curve is from the non-fractionated sample. (From Ref. [132] with permission)

distribution of the sample under investigation. In contrast to this, the point by point composition curve of the sample as constructed from dual detection SEC analysis implied chemical homogeneity.

The reason for this discrepancy can be deduced from Fig. 30: all the constituents differing in chemical composition had approximately equal SEC distributions. Dual detection SEC can provide only information on the average composition of the eluted portions. The broad but symmetric CCD of the sample yield roughly constant values for the average at any point of the SEC curve for the whole sample. The investigation is a fine demonstration for the advantage of cross-fractionation over dual-detection SEC [131, 132].

The chromatographic cross-fractionation of a high-conversion poly(styrene-*co*-methyl acrylate) sample with 56.4% methyl acrylate (MA) units and $\bar{M}_n = 482{,}000$ g/mol was performed by SEC and subsequent TLC of the fractions. A preparative SEC column (600 × 25 mm) was used with chloroform eluent at a flow rate of 5 ml/min. The separation of 10 mg sample (injection volume 2 ml) into 11 fractions was repeated one hundred times. The respective fractions were combined, concentrated in a rotary evaporator, and precipitated by pouring into an excess of methanol. The dried polymer from the SEC fractions was investigated by TLC on precoated silica plates through gradient elution with 6–43 vol % methylacetate in tetrachloromethane. The chromatograms were evaluated quantitatively with a chromatoscanner after carbonizing the polymeric material by sulfuric acid. The average MA content calculated from the TLC evaluation was close to the results obtained from dual detection of the sample [133].

Poly(styrene-*co*-*n*-butyl methacrylate) samples were studied by a method called *orthogonal chromatography*. This term refers to CCF by means of two coupled SEC apparatus [134]. In an advanced version, the first one (GPC 1) was equipped with three μStyragel columns. It was run with THF eluent at a 2 ml/min flow rate. The eluate of GPC 1 flowed through the sample loop of GPC 2. This apparatus contained three columns packed with silica-based ether bonded-phase (μBondagel, d_O = 100, 50, and 12.5 nm, respectively) and was run with mixtures of THF and *n*-heptane at 0.5 ml/min flow rate. Best results were obtained with 63.8% *n*-heptane. The orthogonal chromatography was started with 0.2–0.5 ml of a sample solution containing 0.2 to 0.4% polymer, i.e., with m_O = 0.4–2.0 mg. At every eluate slice of interest, the flow through GPC 1 was interrupted and the content of the sample loop injected into GPC 2. The best result obtained by orthogonal chromatography was the separation of a SMA copolymer with 53.3% styrene units from both parent homopolymers, polystyrene, and poly(*n*-butyl methacrylate) [135]. The separation was performed isocratically. The peaks occurred at elution volumes between the interstitial volume and the void volume of the column, i.e., in the range of SEC separation, but the results were not simply due to a size exclusion mechanism. The authors observed that peak positions were not reproducible enough to establish calibration with respect to copolymer composition [136]. Polystyrene remained sometimes partly on the column and appeared in the next samples chromatogram or was washed off the column on injection of pure THF. The use of a scanning detector enabled a quantitative evaluation of chromatograms from GPC 2.

Recently, chromatographic cross-fractionation of random poly(styrene-*co*-ethyl methacrylate) samples was carried out by the combination of SEC and TLC. The use of chromarods together with a flame ionization detector allowed the quantitative interpretation of the TLC results and the evaluation of the two-dimensional distribution of the investigated samples [137].

6.3 CCF of Poly(styrene-*co*-acrylonitrile)

Random copolymers from styrene (S) and acrylonitrile (AN) were investigated using the combination of SEC and HPPLC.

The SEC was performed with a bank of 5 columns, each 300 × 7.8 mm, packed with polystyrene-gel, $d_P = 10$ μm, nominal pore size 500, 10^3, 10^4, 10^5, and 10^6 Å, respectively. The eluent was THF at a flow rate of 1 ml/min. The volume of the sample loop was 0.5 ml. The HPPLC columns used for CCF were either RP18 (150 × 4.6 mm; $d_O = 6$ nm; $d_P = 10$ μm), or RP18 (150 × 4.6 mm; $d_O = 6$ nm; $d_P = 5$ μm), or CN-bonded phase (150 × 4.6 mm; $d_O = 6$ nm; $d_P = 5$ μm). Gradient elution was carried out with *iso*-octane/tetrahydrofuran (the latter often with an admixture of 10% methanol, as described in Sect. 5).

An example was the CCF of the mixture of two SAN model copolymers which had been prepared under conditions giving narrow CCD and MMD. The copolymer SAN I contained 23 wt % AN and had the molar mass $\overline{M}_n = 480{,}000$ g/mol, SAN II contained 19% AN and had $\overline{M}_n = 49{,}000$ g/mol. A sample of 0.95 mg containing 63.3% SAN I and 36.7% SAN II was SEC fractionated into 6 portions. After evaporation of superfluous THF and adjusting the polymer concentration to 0.02–0.05%, a part of the respective SEC fractions containing about 10 μg sample material was injected into the HPPLC apparatus with the mentioned C 18 column. The result clearly showed a separation of the two copolymers in the sample mixture [138]. Since retention in HPPLC increased with AN content and molar mass under the conditions reported, a more rigorous test was performed using the mixture of 59.3% SAN III and 40.7% SAN IV. The sample SAN III contained 16.1 wt % AN and had the molar mass $\overline{M}_n = 325{,}000$ g/mol, SAN IV contained 30% AN and had $\overline{M}_n = 86{,}000$ g/mol. Thus, the separation of SAN III from SAN IV by composition was rendered difficult through the opposing influence of molar mass. A 0.87 mg sample of the mixture was subdivided into 6 portion by SEC. Although the SEC elugram did not show the presence of two polymers, the HPPLC clearly revealed the two components. Now, the part of the SEC fractionation was repeated where parts of both copolymers SAN III and SAN IV were co-eluted. From eluate slices of 0.5 ml each 100 μl (or 175 μl from the last one) were injected into the HPPLC apparatus without any pretreatment. The result of this cross-fractionation is shown in Fig. 31 [139].

The combination of SEC and HPPLC was also applied to commercial copolymers [140, 141]. Among the prerequisites for this task is the reliable calibration which provides the possibility of evaluating the chemical composition (AN content) of each constituent of a copolymer from its HPPLC retention and molar mass. This requires knowledge of the molar-mass dependence of solubility which can be gained by HPPLC investigation of SEC fractions. Figure 32 shows that a suitable mixture of selected standards can provide many data by only a few experiments. From the average position of the HPPLC peaks, the gradient program, and the lag time of the apparatus (i.e., the break-through time of the gradient) the solvent/nonsolvent composition can be calculated for each fractional constituent of the mixture.

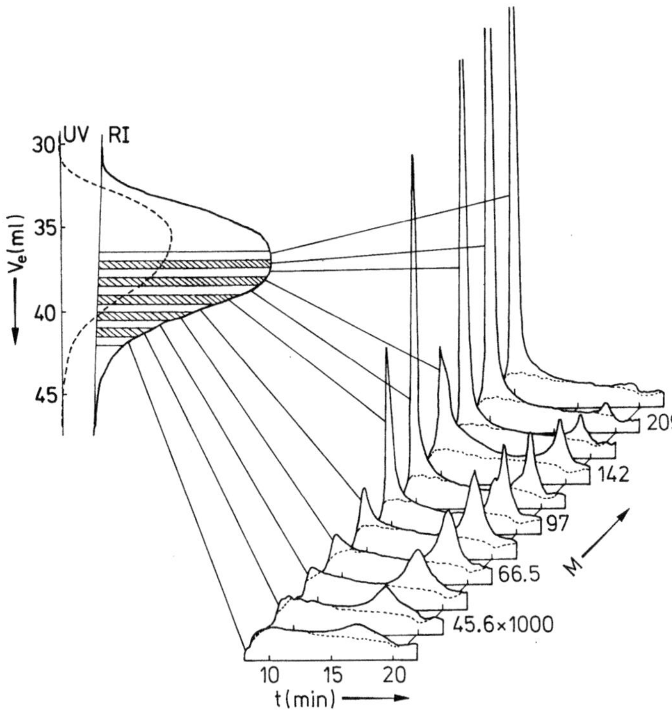

Fig. 31. Chromatographic cross-fractionation of 0.87 mg poly(styrene-*co*-acrylonitrile). The sample was a mixture of 59.3% copolymer with 16.1% AN and 40.7% of another copolymer with 30% AN. The latter produced the band at 15–20 min elution time, the former the band at about 9 min. SEC slices of 0.5 ml each as indicated in the upper right part of the figure. They were used without any additional treatment for the second separation. The molar mass data indicated refer to the shaded SEC slices [139].

The volume fraction of nonsolvent, φ^*, in the eluent delivering a polymer of molar mass M is:

$$\varphi^* = A + B/\sqrt{M} \tag{13}$$

with A and B being constant for a given polymer/solvent system.

Equation (13) had been found empirically by means of turbidimetric titrations [142], but is, to the best of our knowledge, at the same time the valid representation of HPPLC data.

The recent CCF evaluation [141] was performed on the basis of Eq. (13) through:

$$w = 33.12 + 84.9\varphi^*_{(140,w)} \tag{14}$$

with w being the weight percentage of acrylonitrile in a fractional constituent, and $\varphi^*_{(140,w)}$ the volume fraction of nonsolvent in the eluent mixture delivering a sample of the same composition but having the reference value of molar mass 140,000 g/mol [141].

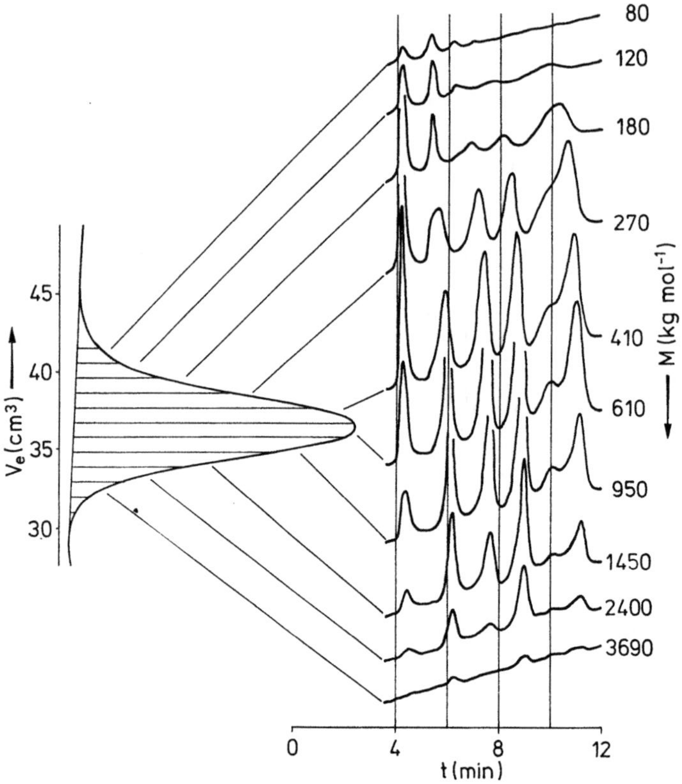

Fig. 32. Evaluation of the molar mass calibration. Investigation of the mixture of five SAN copolymers through SEC and subsequent HPLC on a CN column (150 × 4.6 mm; $d_O = 6$ nm; $d_P = 5$ μm). Gradient: *iso*-octane/THF, from 50 to 100% in 10 min. Samples (in the order of elution): 16.1 wt % AN (325 kg/mol), 23.0% (480), 29.1% (510), 36.4% (380), 42.5% (340). (By courtesy of Elsevier Appl. Sci. Publ., Barking, UK [141])

7 Conclusions and Outlook

Having compiled the results from numerous investigations into the molecular structure of copolymers, the authors from recent reviews concluded "that the composition heterogeneity in copolymers prepared in bulk at high conversions can hardly be predicted within the framework of theories available to date. Therefore, need for improving experimental methods to determine the composition heterogeneity is apparent. In this connection it should be pointed out that the sole use of SEC gives only information on the composition drift (heterogeneity) and not on the CCD, unless a unique correlation exists between composition and chain length for a given copolymerization systems."

"It is the combination of different techniques which is important in the characterization of copolymers for their different heterogeneities."

The author of the present pages agrees completely. It is his conviction that chromatographic techniques can essentially contribute to the solution of many problems involved in copolymer research, not least because of the marvellous acceleration of experimental work. The destination was set just 50 years ago [31]. Now we are in the position to reach this goal. The many problems still existing shall not be denied, but the combined efforts of theorists and experimentators, the use of timely equipment, and, last but not least, the mutual utilization of experience gathered with synthetic polymers or biopolymers, have provided the potential for developing rapid and precise methods that can be used routineously.

8 References

1. Giddings, J. C.: J. Gas Chromatogr. 413 (1967)
2. Thies, C.: J. Polymer Sci. C, Polymer Symposia *34*, 201 (1971)
3. Larmann, J. P., DeStefano, J. J., Goldberg, A. P., Stout, R. W., Snyder, L. R., Stadalius, M. A.: J. Chromatogr. *255*, 163 (1983)
4. Lai, S.-T., Sangermano, L., Locke, D. C.: J. Chromatogr. *312*, 313 (1984)
5. Benedek, K., Dong, S., Karger, B. L.: J. Chromatogr. *317*, 227 (1984)
6. Knox, J. H., McLennan, F.: J. Chromatogr. *185*, 289 (1979)
7. Tennikov, M. B., Nefedov, P. P., Lagareva, M. A., Frenkel, S. Ja.: Vysokomol. Soed. Ser. A *19*, 657 (1977)
8. Belenkii, B. G.: Pure Appl. Chem. *51*, 1519 (1979)
9. Mourey, T. H.: private communication, 1983
10. Balke, S. T., Patel, R. D.: ACS Advances in Chemistry Series *203*, Craver, C. D. (ed), Amer. Chem. Soc. 1983
11. Klein, J., Leidigkeit, G.: Makromol. Chem. *180*, 2753 (1979)
12. Snyder, L. R., Kirkland, J. J.: "Introduction to Modern Liquid Chromatography", 2nd edition, John Wiley & Sons, Inc., New York 1979, Table 16.3
13. Inagaki, H.: Polymer Sci. *24*, 190 (1977)
14. Eisenbeiß, F., Dumont, E., Henke, H.: Angew. Makromol. Chem. *71*, 67 (1978)
15. Engelhardt, H., Müller, H.: Chromatographia *19*, 77 (1984)
16. Mori, S., Suzuki, T.: J. Liquid Chromatogr. *4*, 1685 (1981)
17. Mori, S.: Adv. Chromatography *21*, 187 (1983)
18. Quivoron, C., in: Steric Exclusion Liquid Chromatography of Polymers, Janča, J. (ed), Chromatographic Science Series, Volume 25, pp. 213, Marcel Dekker, New York 1984
19. Glöckner, G.: "Polymer Characterization by Liquid Chromatography", Chapter 19.7, Elsevier, Amsterdam, in press
20. Yau, W. W., Kirkland, J. J., Bly, D. D.: "Modern Size-Exclusion Liquid Chromatography", John Wiley & Sons, New York 1979, Chapter 12.9
21. Inagaki, H., in: "Fractionation of Synthetic Copolymers", Tung, L. H. (ed), Chapter 8, Marcel Dekker, New York 1977
22. Belenkii, B. G., Gankina, E. S.: J. Chromatogr. *141*, 13 (1977)
23. Unger, K.: "Porous Silica — Its Properties and Use as Support in Column Liquid Chromatography", Elsevier, Amsterdam 1979
24. Snyder, L. R., Kirkland, J. J.: Ref. 12, Table 6.1
25. Snyder, L. R.: Principles of Adsorption Chromatography", Marcel Dekker, New York 1968
26. Taga, T., Inagaki, H.: Angew. Makromol. Chem. *33*, 129 (1973)
27. Månsson, P.: J. Polymer Sci. A-1, Polymer Chem. Ed. *18*, 1945 (1980)
28. Andrews, G. D., Vatvars, A.: Macromolecules *14*, 1603 (1981)
29. Winnik, M. A., Paton, K., Danhelka, J., Redpath, A. E. C.: J. Chromatogr. *242*, 97 (1982)
30. Mourey, T. H., Noh, I., Yu, H.: J. Chromatogr. *303*, 361 (1984)
31. Mark, H., Saito, G.: Monatsh. Chemie *68*, 237 (1936)
32. Landler, I.: C. R. hebd. Séances Acad. Sci. (Paris) *225*, 629 (1947)
33. Golub, M. A.: J. Polymer Sci. *11*, 583 (1953)
34. Tagata, N., Homma, T., Nippon Kagaku Kaishi (J. Chem. Soc. Japan, Chem. Ind. Chem.) 7, 1330 (1972)

35. Horii, F., Ikada, Y.: J. Polymer Sci. B, Polymer Letters *12*, 27 (1974)
36. Wesslen, B., Månsson, P.: J. Polymer Sci. A-1, Polymer Chem. *13*, 2545 (1975)
37. Teramachi, S., Esaki, H.: Polymer J. *7*, 593 (1975)
38. Teramachi, S., Hasegawa, A., Shima, Y., Akatsuka, M., Nakajima, M.: Macromolecules *12*, 992 (1979)
39. Sato, H., Takeuchi, H., Suzuki, S., Tanaka, Y.: Makromol. Chem., Rapid Commun. *5*, 719 (1984)
40. Mori, S.: Polymer 85, Melbourne, Australia, Paper No 037,94
41. Hara, S., Ohkuma, T.: J. Chromatogr. *316*, 389 (1984)
42. Danielewicz, M., Kubin, M.: J. Appl. Polymer Sci. *26*, 951 (1981)
43. Glöckner, G., van den Berg, J. H. M.: 9th International Symposium on Column Liquid Chromatography, Edinburgh, Contribution PO2.15
44. Teramachi, S.: private communication, 1984
45. Tanaka, Y., Sato, H., Takeuchi, H.: Polymer Preprints, Japan *33*, 752 (1984)
46. Mourey, T. H., Smith, G. A., Snyder, L. R.: Anal. Chem. *56*, 1773 (1984)
47. Mourey, T. H.: Anal. Chem. *56*, 1777 (1984)
48. Lai, S.-T., Locke, D.: J. Chromatogr. *252*, 325 (1982)
49. Schmitz, F. P., Hilgers, H., Klesper, E.: J. Chromatogr. *267*, 267 (1983)
50. Howard, G. A., Martin, A. J. P.: Biochem. J. *46*, 532 (1950)
51. Verzele, M., Mussche, P., Sandra: J. Chromatogr. *190*, 331 (1980)
52. Fazio, S., Crowther, J., Meyer, R., Tomellini, S., Floyd, T., Hartwick, R.: 8th International Symposium on Column Liquid Chromatography, New York, Contribution 1a-B2
53. Bayer, E., Albert, K., Reiners, J., Nieder, M., Müller, D.: J. Chromatogr. *264*, 197 (1983)
54. Horváth, Cs.: Liq. Chromatogr. *1*, 552 (1983)
55. Melander, W. R., Corradini, D., Horváth, Cs.: J. Chromatogr. *317*, 67 (1984)
56. Miller, N. T., Feibush, B., Karger, B. L.: J. Chromatogr. *316*, 519 (1984)
57. Smolenski, K. A., Fallon, A., Light, N.: J. Chromatogr. *287*, 29 (1984)
58. Doris, P. A.: J. Chromatogr. *336*, 392 (1984)
59. Luiken, J., van der Zee, R., Welling, G. W.: J. Chromatogr. *284*, 482 (1984)
60. Herskovits, T. T., Gadegbeku, B., Jaillet, H.: J. Biol. Chem. *245*, 2588 (1970)
61. Wilks, J. H., Butler, S. S.: J. Chromatogr. *298*, 123 (1984)
62. Barford, R. A., Sliwinski, B. J., Breyer, A. C., Rothbart, H. L.: J. Chromatogr. *235*, 281 (1982)
63. Chang, J. P.: J. Chromatogr. *317*, 157 (1984)
64. Shaltiel, S.: Methods Enzymol. *104*, 69 (1984)
65. Emanuel, R. L., Williams, G. H., Giese, R. W.: J. Chromatogr. *312*, 285 (1984)
66. Crabb, J. W., Heilmeyer, L. M. G.: J. Chromatogr. *296*, 129 (1984)
67. Phelan, M. A., Cohen, K. A.: J. Chromatogr. *266*, 55 (1983)
68. Gurley, L. R., Prentice, D. A., Valdez, J. G., Spall, W. D.: J. Chromatogr. *266*, 609 (1983)
69. Taneja, A. K., Lau, S. Y. M., Hodges, R. S.: J. Chromatogr. *317*, 1 (1984)
70. Skinner, S. J. M., Grego, B., Hearn, M. T. W., Liggins, G. C.: J. Chromatogr. *308*, 111 (1984)
71. Heukeshoven, J., Dernick, R.: Chromatographia *19*, 95 (1984)
72. Regnier, F. E.: Methods Enzymol. *91*, 137 (1983)
73. Moore, R. M., Walters, R. R.: J. Chromatogr. *317*, 119 (1984)
74. El Rassi, Z., Horváth, Cs.: Chromatographia *19*, 1 (1984)
75. Laurent, C. J. C. M., Billiet, H. A. H., De Galan, L., Buytenhuys, F. A., van der Maeden, F. P. B.: J. Chromatogr. *287*, 45 (1984)
76. Chaiken, I. M., Miller, T. H., Sequeira, R. P., Kanmera, T.: Analytical Biochemistry *143*, 215 (1984)
77. O'Keefe, J. H., Sharry, L. F., Jones, A. J.: J. Chromatogr. *336*, 73 (1984)
78. Welinder, B. S., Linde, S., Hansen, B.: J. Chromatogr. *265*, 301 (1983)
79. Halperin, G., Tauber-Finkelstein, M., Shaltiel, S.: J. Chromatogr. *317*, 103 (1984)
80. Kamp, R. M., Bosserhoff, A., Kamp, D., Wittmann-Liebold, B.: J. Chromatogr. *317*, 181 (1984)
81. Kerlavage, A. R., Weitzmann, C. J., Cooperman, B. S.: J. Chromatogr. *317*, 201 (1984)
82. Wells, R. D.: J. Chromatogr. *336*, 3 (1984)
83. Bischoff, R., McLaughlin, L. W.: J. Chromatogr. *296*, 329 (1984)
84. Kirkland, J. J.: Chromatographia *8*, 661 (1975)
85. Parris, N. A.: J. Chromatogr. *157*, 161 (1978)

86. Lattimer, R. P., Harmon, D. J., Welch, K. R.: Anal. Chem. *51*, 1293 (1979)
87. Melander, W. R., Nahum, A., Horváth, Cs.: J. Chromatogr. *185*, 129 (1979)
88. Andrews, G. D., Vatvars, A., Pruckmayr, G.: Macromolecules *15*, 1580 (1982)
89. Hagnauer, G. L.: Polymer Composites *1*, 81 (1980)
90. Hagnauer, G. L.: 35th Pittsburgh Conference on Analytical Chemistry and Spectroscopy, 1984, Contribution No 1984
91. Bij, K. E., Horváth, Cs., Melander, W. R., Nahum, A.: J. Chromatogr. *203*, 65 (1981)
92. Locke, D. C.: J. Chromatogr. Sci. *12*, 433 (1974)
93. Karger, B. L., Gant, J. R., Hartkopf, A., Weinert, P. H.: J. Chromatogr. *128*, 65 (1976)
94. Hennion, M. C., Picard, C., Combellas, C., Caude, M., Rosset, R.: J. Chromatogr. *210*, 211 (1981)
95. Baker, C. A., Williams, R. P. J.: J. Chem. Soc. (London) 2352 (1956)
96. Barrall, E. M., Johnson, J. F., Cooper, A. R.: "Column Fractionation", in: Tung, L. H. (ed): "Fractionation of Synthetic Polymers, Marcel Dekker Inc., New York 1977
97. Cooper, A. R.: "Column Fractionation of Macromolecules by Gradient Elution", in: Epton, R. (ed): "Chromatography of Synthetic and Biological Polymers", Vol. 1, Ellis Horwood Ltd. Publ., Chichester, U.K. 1978
98. Cooper, W., Vaughan, G., Madden, R. W.: J. Appl. Polymer Sci. *1*, 329 (1959)
99. Acres, G. J. K., Dalton, F. L.: J. Polymer Sci. A, General Papers *1*, 2419 (1963)
100. Uraneck, C. A., Barker, M. G., Johnson, W. D.: Rubber Chem. & Technol. *38*, 802 (1965)
101. Ding, H.-C., Pan, H.-C., Yu, T.-K.: Gaofenzi Tongxun (China) *7*, 260 (1965)
102. Blass, E., Seide, H.: Plaste & Kautschuk *13*, 405 (1966)
103. Danon, J., Jozefonvicz: Europ. Polymer J. *5*, 405 (1969)
104. Braun, D., Chaudhari, D.: Makromol. Chem. *133*, 241 (1970)
105. Florián, S., Lath, D., Ďurďovič, V.: Angew. Makromol. Chem. *10*, 189 (1970)
106. Kálal, J., Marousek, V., Švec, F.: Angew. Makromol. Chem. *38*, 45 (1974)
107. Ogawa, T., Inaba, T.: J. Polymer Sci. A-2, Polymer Phys. *12*, 785 (1974)
108. Lovrić, L.: J. Polymer Sci. A-2, Polymer Phys. *7*, 1357 (1969)
109. Glöckner, G., Francuskiewicz, F., Müller, K.-D.: Plaste & Kautschuk *18*, 654 (1971)
110. Scholtan, W., Kwoll, F. J.: Makromol. Chem. *151*, 33 (1972)
111. Ogawa, T., Sakai, M.: J. Polymer Sci. A-2, Polymer Phys. *19*, 1377 (1981)
112. Teramachi, S., Fukao, T.: Polymer J. *6*, 532 (1974)
113. Teramachi, S., Obata, R., Yamashita, K., Takemoto, N.: J. Makromol. Sci. A *11*, 535 (1977)
114. Glöckner, G., Kroschwitz, H., Meissner, Ch.: Acta Polymerica *33*, 614 (1982)
115. Danielewicz, M., Kubin, M., Vozka, S.: J. Appl. Polymer Sci. *27*, 3629 (1982)
116. Armstrong, D. W., Bui, K. H.: Anal. Chem. *54*, 706 (1982)
117. Boehm, R. E., Martire, D. E., Armstrong, D. W., Bui, K. H.: Macromolecules *16*, 466 (1983)
118. Armstrong, D. W., Bui, K. H., Boehm, R. E.: J. Liquid Chromatogr. *6*, 1 (1983)
119. Glöckner, G.: Pure & Appl. Chem. *55*, 1553 (1983)
120. Armstrong, D. W., Boehm, R. E.: J. Chromatogr. Sci. *22*, 378 (1984)
121. Quarry, M. A., Stadalius, M. A., Mourey, T. H., Snyder, L. R.: in preparation
122. Stadalius, M. A., Quarry, M. A., Mourey, T. H., Snyder, L. R.: in preparation
123. Glöckner, G., van den Berg, J. H. M.: Chromatographia *19*, 55 (1984)
124. Glöckner, G., van den Berg, J. H. M.: 5th Danube Symposium on Chromatography, Yalta (USSR) 1985, Contribution N° 067 (S. 14-1)
125. Topčiev, A. V., Litmanovič, A. D., Štern, V. Ja.: Dokl. Akad. Nauk SSSR *147*, 1389 (1962)
126. Johnson, E. L., Gloor, R., Majors, R. E.: J. Chromatogr. *149*, 571 (1978)
127. Ogawa, T., Tanaka, S., Inaba, T.: J. Appl. Polymer Sci. *17*, 319 (1973)
128. Nakano, S., Goto, Y.: J. Appl. Polymer Sci. *26*, 4217 (1981)
129. Wild, L., Ryle, T. R., Knobeloch, D. C.: Polymer Prepr., Am. Chem. Soc., Polym. Chem. Div. *23*, 2, 133 (1982)
130. Belenkii, B. G., Gankina, E. S., Nefedov, P. P., Lazareva, M. A., Savitskaya, T. S., Volchikhina, M. D.: J. Chromatogr. *108*, 61 (1975)
131. Tanaka, M., Omoto, M., Donkai, N., Inagaki, H.: J. Macromol. Sci. — Phys., B *17*, 211 (1980)
132. Inagaki, H., Tanaka, T.: Pure & Appl. Chem. *54*, 309 (1982)
133. Teramachi, S., Hasegawa, A., Yoshida, S.: Macromolecules *16*, 542 (1983)
134. Balke, S. T., Patel, R. D.: J. Polymer Sci. B, Polymer Letters, *18*, 453 (1980)

135. Balke, S. T.: Separation and Purification Methods *11*, 1 (1982)
136. Balke, S. T., Patel, R. D.: ACS Advances in Chemistry Series *203*, 281 (1983)
137. Tacx, J. C. J. F., Ammerderffer, J. L., German, A. L.: International Discussion Meeting on Polymer Science and Technology, Rolduc Abbey, The Netherlands 1985
138. Glöckner, G., van den Berg, J. H. M., Meijerink, N. L., Scholte, T. G., Koningsveld, R.: Macromolecules *17*, 962 (1984)
139. Glöckner, G., Koningsveld, R.: Makromol. Chem., Rapid Commun. *4*, 529 (1983)
140. Glöckner, G., van den Berg, J. H. M., Meijerink, N. L., Scholte, T. G., Koningsveld, R.: J. Chromatogr. *317*, 615 (1984)
141. Glöckner, G., van den Berg, J. H. M., Meijerink, N. L., Scholte, T. G.: "Characterization of copolymers: chromatographic cross-fractionation analysis of styrene-acrylonitrile copolymers", in: Kleintjens, L., Lemstra, P. (ed): "Integration of Fundamental Polymer Science and Technology", Elsevier Applied Science Publ., Barking, UK (1986), p. 85
142. Glöckner, G.: Z. physik. Chemie (Leipzig) *229*, 98 (1965)

Editor: G. Heublein
Received January 19, 1986

Author Index Volumes 1–79

Subject Index

GPSR Compliance

The European Union's (EU) General Product Safety Regulation (GPSR) is a set of rules that requires consumer products to be safe and our obligations to ensure this.

If you have any concerns about our products, you can contact us on ProductSafety@springernature.com

In case Publisher is established outside the EU, the EU authorized representative is:

Springer Nature Customer Service Center GmbH
Europaplatz 3
69115 Heidelberg, Germany

Batch number: 10620881

Printed by Printforce, the Netherlands